PRAISE FOR R

D0441347

The IMMORTAL LIFE

Winner of 2010 *Chicago Tribune* Heartland Prize for Nonfiction

Winner of 2010 Wellcome Trust Book Prize

New York Times Notable Book
New Yorker Reviewers' Favorite
Entertainment Weekly Best Book of the Year
O, The Oprah Magazine Best Book of the Year
National Public Radio Best of the Bestsellers
Financial Times Nonfiction Favorite
Los Angeles Times Critics' Pick
New York magazine Top Ten Book of the Year
Discover magazine 2010 Must-Read
Independent (U.K.) Best Book of the Year
Publishers Weekly Best Book of the Year
Library Journal Best Book of the Year
Kirkus Reviews Best Book of the Year
Times (U.K.) Best Book of the Year
U.S. News & World Report Talk-Worthy Title
Booklist Best Book of the Year
Globe and Mail Best Book of the Year

"I could not put the book down . . . The story of modern medicine and bioethics—and, indeed, race relations—is refracted beautifully, and movingly." **—ENTERTAINMENT WEEKLY**

"Science writing is often just about 'the facts.' Skloot's book, her first, is far deeper, braver, and more wonderful." **—NEW YORK TIMES BOOK REVIEW**

"*The Immortal Life of Henrietta Lacks* is a triumph of science writing . . . one of the best nonfiction books I have ever read." **—WIRED.COM**

"A deftly crafted investigation of a social wrong committed by the medical establishment, as well as the scientific and medical miracles to which it led." **—WASHINGTON POST**

"Riveting . . . a tour-de-force debut." **—CHICAGO SUN-TIMES**

"A real-life detective story, *The Immortal Life of Henrietta Lacks* probes deeply into racial and ethical issues in medicine . . . The emotional impact of Skloot's tale is intensified by its skillfully orchestrated counterpoint between two worlds." **—NATURE**

"Read this . . . By letting the Lackses be people, and by putting them in the center of the history, Skloot turns just another tale about the march of progress into a complicated portrait of the interaction between science and human lives. **—BOINGBOING.NET**

"[A] remarkable and moving book . . . a vivid portrait of Lacks that should be as abiding as her cells." **—THE TIMES** (U.K.)

"I can't imagine a better tale. A detective story that's at once mythically large and painfully intimate. I highly recommend this book."
—Jad Abumrad, **RADIOLAB**

"Skloot is a terrific popularizer of medical science, guiding readers through this dense material with a light and entertaining touch."
—THE GLOBE AND MAIL (Canada)

"A rare and powerful combination of race, class, gender, medicine, bioethics, and intellectual property; far more rare is the writer that can so clearly fuse those disparate threads into a personal story so rich and compelling."
—SEED

"Powerful story . . . I feel moved even to say on behalf of the thousands of anonymous black men and women who've been experimented on for medical purposes, thank you. Thank you for writing this important book."
—Kali-Ahset Amen, **RADIO DIASPORA**

"Skloot has written an important work of immersive nonfiction that brings not only the stories of Henrietta Lacks and HeLa once more into line, but also catharsis to a family in sore need of it."
—THE TIMES LITERARY SUPPLEMENT

"A masterful work of nonfiction . . . a real page turner."
—Hanna Rosin, **SLATE**

"Skloot explores human consequences of the intersection of science and business, rescuing one of modern medicine's inadvertent pioneers from an unmarked grave." **—U.S. NEWS & WORLD REPORT**

"Remarkably balanced and nonjudgmental . . . *The Immortal Life of Henrietta Lacks* will leave readers reeling, plain and simple. It has a power and resonance rarely found in any genre, and is a subject that touches each of us, whether or not we are aware of our connection to Henrietta's gift."
—THE OREGONIAN

"This is the perfect book. It reads like a novel but has the intellectual substance of a science textbook or a historical biography."
—THE DAILY NEBRASKAN

"Illuminates what happens when medical research is conducted within an unequal health-care system and delivers an American narrative fraught with intrigue, tragedy, triumph, pathos, and redemption." **—MS.**

"A tremendous accomplishment—a tale of important science history that reads like a terrific novel." **—KANSAS CITY STAR**

"Good science writing isn't easy, but Skloot makes it appear so." **—THE WICHITA EAGLE**

"Encompasses nearly every hot-button issue currently surrounding the practice of medicine." **—MADISON CAPITAL TIMES**

"Defies easy categorization . . . as unpredictable as any pulp mystery and as strange as any science fiction." **—WILLAMETTE WEEK**

"An achievement . . . navigates both the technical and deeply personal sides of the HeLa story with clarity and care." **—THE PORTLAND MERCURY**

"[A] remarkable book." **—LONDON REVIEW OF BOOKS**

"An essential reminder that all human cells grown in labs across the world, HeLa or otherwise, came from individuals with fears, desires, and stories to tell." **—CHEMICAL & ENGINEERING NEWS**

"Blows away the notion that science writing must be the literary equivalent to Ambien." **—CHICAGO TRIBUNE**

"Seldom do you read a book that is science, social history, and a page turner." **—BRITISH MEDICAL JOURNAL**

"Thrilling and original nonfiction that refuses to be shoehorned into anything as trivial as a genre. It is equal parts popular science, historical biography, and detective novel." **—Ed Yong, DISCOVER.COM**

"Best book I've read in years." **—Brian Sullivan, FOX BUSINESS NETWORK**

"Thanks to Rebecca Skloot, we may now remember Henrietta—who she was, how she lived, how she died." **—THE NEW REPUBLIC**

"We need more writers like Rebecca Skloot." **—E. O. WILSON**

The
IMMORTAL
LIFE
of
HENRIETTA
LACKS

Rebecca
Skloot

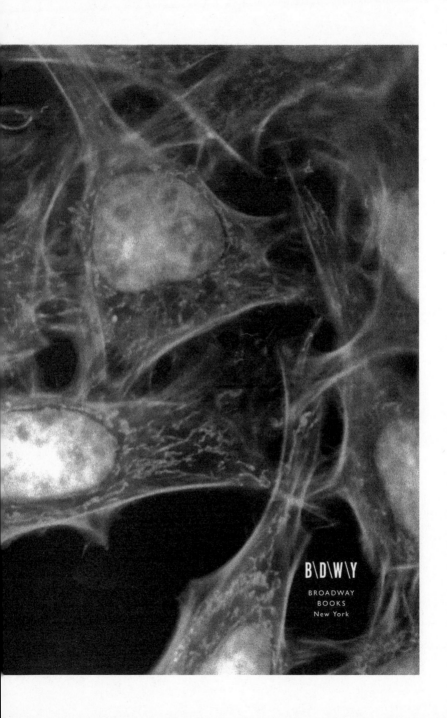

B\D\W\Y

BROADWAY
BOOKS
New York

Published in the United States by Broadway Books,
an imprint of the Crown Publishing Group,
a division of Penguin Random House LLC, New York.
broadwaybooks.com

BROADWAY BOOKS and its logo, B \ D \ W \ Y,
are trademarks of Penguin Random House LLC.

Originally published in hardcover in slightly different form in
the United States by Crown, an imprint of the
Crown Publishing Group, a division of
Penguin Random House LLC, New York, in 2010.

Library of Congress Cataloging-in-Publication Data
Skloot, Rebecca, 1972–
    The immortal life of Henrietta Lacks / Rebecca Skloot.
        p. cm.
    Includes bibliographical references and index.
    1. Lacks, Henrietta, 1920–1951—Health. 2. Cancer—Patients—
Virginia—Biography. 3. African American women—History.
4. Human experimentation in medicine—United States—History.
5. HeLa cells. 6. Cancer—Research. 7. Cell culture. 8. Medical
ethics. I. Title.
    RC265.6.L24S55   2009
    616'.02774092—dc22
    [B]   2009031785

ISBN 978-0-8041-9010-7
EBOOK ISBN 978-0-307-58938-5

Printed in the United States of America

DESIGN BY BARBARA STURMAN
COVER ART: © 2017 HOME BOX OFFICE, INC.
ALL RIGHTS RESERVED. HBO® IS A SERVICE MARK OF
HOME BOX OFFICE, INC.

Photographs on pages vi-vii, 11, 87, and 177
copyright © 2010 Omar A. Quintero.
Photograph on page xviii courtesy of the Lacks Family,
via The Henrietta Lacks Foundation.

10 9 8 7 6 5 4 3 2 1

First Movie Tie-in Paperback Edition

For my family:
My parents, Betsy and Floyd; their spouses, Terry and Beverly;
my brother and sister-in-law, Matt and Renee;
and my wonderful nephews, Nick and Justin.
They all did without me for far too long because of this book,
but never stopped believing in it, or me.

And in loving memory of my grandfather,
James Robert Lee (1912–2003),
who treasured books more than anyone I've known.

Contents

Part Three

IMMORTALITY

A Few Words About This Book

This is a work of nonfiction. No names have been changed, no characters invented, no events fabricated. While writing this book, I conducted more than a thousand hours of interviews with family and friends of Henrietta Lacks, as well as with lawyers, ethicists, scientists, and journalists who've written about the Lacks family. I also relied on extensive archival photos and documents, scientific and historical research, and the personal journals of Henrietta's daughter, Deborah Lacks.

I've done my best to capture the language with which each person spoke and wrote: dialogue appears in native dialects; passages from diaries and other personal writings are quoted exactly as written. As one of Henrietta's relatives said to me, "If you pretty up how people spoke and change the things they said, that's dishonest. It's taking away their lives, their experiences, and their selves." In many places I've adopted the words interviewees used to describe their worlds and experiences. In doing so, I've used the language of their times and backgrounds, including words such as *colored*. Members of the Lacks family often referred to Johns Hopkins as "John Hopkin," and I've kept their usage when they're speaking. Anything written in the first person in Deborah Lacks's voice is a quote of her speaking, edited for length and occasionally clarity.

Since Henrietta Lacks died decades before I began writing this book, I relied on interviews, legal documents, and her medical records to re-create scenes from her life. In those scenes, dialogue is either deduced from the written record or quoted verbatim as it was recounted to me in an interview. Whenever possible I conducted multiple interviews with multiple sources to ensure accuracy. The extract from Henrietta's medical record in chapter 1 is a summary of many disparate notations.

The word *HeLa,* used to refer to the cells grown from Henrietta Lacks's cervix, occurs throughout the book. It is pronounced *hee-lah.*

About chronology: Dates for scientific research refer to when the research was conducted, not when it was published. In some cases those dates are approximate because there is no record of exact start dates. Also, because I move back and forth between multiple stories, and scientific discoveries occur over many years, there are places in the book where, for the sake of clarity, I describe scientific discoveries sequentially, even though they took place during the same general period of time.

The history of Henrietta Lacks and the HeLa cells raises important issues regarding science, ethics, race, and class; I've done my best to present them clearly within the narrative of the Lacks story, and I've included an afterword addressing the current legal and ethical debate surrounding tissue ownership and research. There is much more to say on all the issues, but that is beyond the scope of this book, so I will leave it for scholars and experts in the field to address. I hope readers will forgive any omissions.

The
IMMORTAL
LIFE
of
HENRIETTA
LACKS

We must not see *any* person as an abstraction.
Instead, we must see in every person a universe with its own secrets,
with its own treasures, with its own sources of anguish,
and with some measure of triumph.

—Elie Wiesel
from *The Nazi Doctors and the Nuremberg Code*

PROLOGUE

The Woman in the Photograph

There's a photo on my wall of a woman I've never met, its left corner torn and patched together with tape. She looks straight into the camera and smiles, hands on hips, dress suit neatly pressed, lips painted deep red. It's the late 1940s and she hasn't yet reached the age of thirty. Her light brown skin is smooth, her eyes still young and playful, oblivious to the tumor growing inside her—a tumor that would leave her five children motherless and change the future of medicine. Beneath the photo, a caption says her name is "Henrietta Lacks, Helen Lane or Helen Larson."

No one knows who took that picture, but it's appeared hundreds of times in magazines and science textbooks, on blogs and laboratory walls. She's usually identified as Helen Lane, but often she has no name at all. She's simply called HeLa, the code name given to the world's first immortal human cells—*her* cells, cut from her cervix just months before she died.

Her real name is Henrietta Lacks.

I've spent years staring at that photo, wondering what kind of life she led, what happened to her children, and what she'd think about cells from her cervix living on forever—bought, sold, packaged, and shipped by the trillions to laboratories around the world. I've tried to imagine how she'd feel knowing that her cells went up in the first space missions to see what would happen to human cells in zero gravity, or that they helped with some of the most important advances in medicine: the polio vaccine, chemotherapy, cloning, gene mapping, in vitro fertilization. I'm pretty sure that she—like most of us—would be shocked to hear that there are trillions more of her cells growing in laboratories now than there ever were in her body.

There's no way of knowing exactly how many of Henrietta's cells are alive today. One scientist estimates that if you could pile all HeLa cells ever grown onto a scale, they'd weigh more than 50 million metric tons—an inconceivable number, given that an individual cell weighs almost nothing. Another scientist calculated that if you could lay all HeLa cells ever grown end-to-end, they'd wrap around the Earth at least three times, spanning more than 350 million feet. In her prime, Henrietta herself stood only a bit over five feet tall.

I first learned about HeLa cells and the woman behind them in 1988, thirty-seven years after her death, when I was sixteen and sitting in a community college biology class. My instructor, Donald Defler, a gnomish balding man, paced at the front of the lecture hall and flipped on an overhead projector. He pointed to two diagrams that appeared on the wall behind him. They were schematics of the cell reproduction cycle, but to me they just looked like a neon-colored mess of arrows, squares, and circles with words I didn't understand, like "MPF Triggering a Chain Reaction of Protein Activations."

I was a kid who'd failed freshman year at the regular public high school because she never showed up. I'd transferred to an alternative school that offered dream studies instead of biology, so I was taking Defler's class for high-school credit, which meant that I was sitting in a college lecture hall at sixteen with words like *mitosis* and *kinase inhibitors* flying around. I was completely lost.

"Do we have to memorize everything on those diagrams?" one student yelled.

Yes, Defler said, we had to memorize the diagrams, and yes, they'd be on the test, but that didn't matter right then. What he wanted us to understand was that cells are amazing things: There are about one hundred trillion of them in our bodies, each so small that several thousand could fit on the period at the end of this sentence. They make up all our tissues—muscle, bone, blood—which in turn make up our organs.

Under the microscope, a cell looks a lot like a fried egg: It has a white (the *cytoplasm*) that's full of water and proteins to keep it fed, and a yolk (the *nucleus*) that holds all the genetic information that makes you *you*. The cytoplasm buzzes like a New York City street. It's crammed full of molecules and vessels endlessly shuttling enzymes and sugars from one part of the cell to another, pumping water, nutrients, and oxygen in and out of the cell. All the while, little cytoplasmic factories work 24/7, cranking out sugars, fats, proteins, and energy to keep the whole thing running and feed the nucleus—the brains of the operation. Inside every nucleus within each cell in your body, there's an identical copy of your entire genome. That genome tells cells when to grow and divide and makes sure they do their jobs, whether that's controlling your heartbeat or helping your brain understand the words on this page.

Defler paced the front of the classroom telling us how mitosis— the process of cell division—makes it possible for embryos to grow into babies, and for our bodies to create new cells for healing wounds or replenishing blood we've lost. It was beautiful, he said, like a perfectly choreographed dance.

All it takes is one small mistake anywhere in the division process for cells to start growing out of control, he told us. Just *one* enzyme misfiring, just *one* wrong protein activation, and you could have cancer. Mitosis goes haywire, which is how it spreads.

"We learned that by studying cancer cells in culture," Defler said. He grinned and spun to face the board, where he wrote two words in enormous print: HENRIETTA LACKS.

Henrietta died in 1951 from a vicious case of cervical cancer, he told us. But before she died, a surgeon took samples of her tumor and put them in a petri dish. Scientists had been trying to keep human cells alive in culture for decades, but they all eventually died. Henrietta's were different: they reproduced an entire generation every twenty-four hours, and they never stopped. They became the first immortal human cells ever grown in a laboratory.

"Henrietta's cells have now been living outside her body far longer than they ever lived inside it," Defler said. If we went to almost any cell culture lab in the world and opened its freezers, he told us, we'd probably find millions—if not billions—of Henrietta's cells in small vials on ice.

Her cells were part of research into the genes that cause cancer and those that suppress it; they helped develop drugs for treating herpes, leukemia, influenza, hemophilia, and Parkinson's disease; and they've been used to study lactose digestion, sexually transmitted diseases, appendicitis, human longevity, mosquito mating, and the negative cellular effects of working in sewers. Their chromosomes and proteins have been studied with such detail and precision that scientists know their every quirk. Like guinea pigs and mice, Henrietta's cells have become the standard laboratory workhorse.

"HeLa cells were one of the most important things that happened to medicine in the last hundred years," Defler said.

Then, matter-of-factly, almost as an afterthought, he said, "She was a black woman." He erased her name in one fast swipe and blew the chalk from his hands. Class was over.

As the other students filed out of the room, I sat thinking, *That's it? That's all we get? There has to be more to the story.*

I followed Defler to his office.

"Where was she from?" I asked. "Did she know how important her cells were? Did she have any children?"

"I wish I could tell you," he said, "but no one knows anything about her."

After class, I ran home and threw myself onto my bed with my biology textbook. I looked up "cell culture" in the index, and there she was, a small parenthetical:

> In culture, cancer cells can go on dividing indefinitely, if they have a continual supply of nutrients, and thus are said to be "immortal." A striking example is a cell line that has been reproducing in culture since 1951. (Cells of this line are called HeLa cells because their original source was a tumor removed from a woman named Henrietta Lacks.)

That was it. I looked up HeLa in my parents' encyclopedia, then my dictionary: No Henrietta.

As I graduated from high school and worked my way through college toward a biology degree, HeLa cells were omnipresent. I heard about them in histology, neurology, pathology; I used them in experiments on how neighboring cells communicate. But after Mr. Defler, no one mentioned Henrietta.

When I got my first computer in the mid-nineties and started using the Internet, I searched for information about her, but found only confused snippets: most sites said her name was Helen Lane; some said she died in the thirties; others said the forties, fifties, or even sixties. Some said ovarian cancer killed her, others said breast or cervical cancer.

Eventually I tracked down a few magazine articles about her from the seventies. *Ebony* quoted Henrietta's husband saying, "All I remember is that she had this disease, and right after she died they called me in the office wanting to get my permission to take a sample of some kind. I decided not to let them." *Jet* said the family was angry— angry that Henrietta's cells were being sold for twenty-five dollars a vial, and angry that articles had been published about the cells without their knowledge. It said, "Pounding in the back of their heads was a gnawing feeling that science and the press had taken advantage of them."

The articles all ran photos of Henrietta's family: her oldest son sitting at his dining room table in Baltimore, looking at a genetics textbook. Her middle son in military uniform, smiling and holding a baby. But one picture stood out more than any other: in it, Henrietta's daughter, Deborah Lacks, is surrounded by family, everyone smiling, arms around each other, eyes bright and excited. Except Deborah. She stands in the foreground looking alone, almost as if someone pasted her into the photo after the fact. She's twenty-six years old and beautiful, with short brown hair and catlike eyes. But those eyes glare at the camera, hard and serious. The caption said the family had found out just a few months earlier that Henrietta's cells were still alive, yet at that point she'd been dead for twenty-five years.

All of the stories mentioned that scientists had begun doing research on Henrietta's children, but the Lackses didn't seem to know what that research was for. They said they were being tested to see if they had the cancer that killed Henrietta, but according to the reporters, scientists were studying the Lacks family to learn more about Henrietta's cells. The stories quoted her son Lawrence, who wanted to know if the immortality of his mother's cells meant that he might live forever too. But one member of the family remained voiceless: Henrietta's daughter, Deborah.

As I worked my way through graduate school studying writing, I became fixated on the idea of someday telling Henrietta's story. At one point I even called directory assistance in Baltimore looking for Henrietta's husband, David Lacks, but he wasn't listed. I had the idea that I'd write a book that was a biography of both the cells and the woman they came from—someone's daughter, wife, and mother.

I couldn't have imagined it then, but that phone call would mark the beginning of a decadelong adventure through scientific laboratories, hospitals, and mental institutions, with a cast of characters that would include Nobel laureates, grocery store clerks, convicted felons, and a professional con artist. While trying to make sense of the history of cell culture and the complicated ethical debate surrounding the use of human tissues in research, I'd be accused of conspiracy and slammed into a wall both physically and metaphorically, and I'd eventually find myself on the receiving end of something that looked a lot like an exorcism. I did eventually meet Deborah, who would turn out to be one of the strongest and most resilient women I'd ever known. We'd form a deep personal bond, and slowly, without realizing it, I'd become a character in her story, and she in mine.

Deborah and I came from very different cultures: I grew up white and agnostic in the Pacific Northwest, my roots half New York Jew and half Midwestern Protestant; Deborah was a deeply religious black Christian from the South. I tended to leave the room when religion came up in conversation because it made me uncomfortable; Deborah's family tended toward preaching, faith healings, and sometimes voodoo. She grew up in a black neighborhood that was one of the poorest and most dangerous in the country; I grew up in a safe, quiet middle-class neighborhood in a predominantly white city and went to high school with a total of two black students. I was a science journalist who referred to all things supernatural as "woo-woo stuff"; Deborah believed Henrietta's spirit lived on in her cells, controlling the life of anyone who crossed its path. Including me.

"How else do you explain why your science teacher knew her real name when everyone else called her Helen Lane?" Deborah would say. "She was trying to get your attention." This thinking would apply to everything in my life: when I married while writing this book, it was because Henrietta wanted someone to take care of me while I worked. When I divorced, it was because she'd decided he was getting in the way of the book. When an editor who insisted I take the Lacks family out of the book was injured in a mysterious accident, Deborah said that's what happens when you piss Henrietta off.

The Lackses challenged everything I thought I knew about faith, science, journalism, and race. Ultimately, this book is the result. It's not only the story of HeLa cells and Henrietta Lacks, but of Henrietta's family—particularly Deborah—and their lifelong struggle to make peace with the existence of those cells, and the science that made them possible.

Deborah's Voice

When people ask—and seems like people always be askin to where I can't never get away from it—I say, Yeah, that's right, my mother name was Henrietta Lacks, she died in 1951, John Hopkins took her cells and them cells are still livin today, still multiplyin, still growin and spreadin if you don't keep em frozen. Science calls her HeLa and she's all over the world in medical facilities, in all the computers and the Internet everywhere.

When I go to the doctor for my checkups I always say my mother was HeLa. They get all excited, tell me stuff like how her cells helped make my blood pressure medicines and antidepression pills and how all this important stuff in science happen cause of her. But they don't never explain more than just sayin, Yeah, your mother was on the moon, she been in nuclear bombs and made that polio vaccine. I really don't know how she did all that, but I guess I'm glad she did, cause that mean she helpin lots of people. I think she would like that.

But I always have thought it was strange, if our mother cells done so much for medicine, how come her family can't afford to see no doctors? Don't make no sense. People got rich off my mother without us even knowin about them takin her cells, now we don't get a dime. I used to get so mad about that to where it made me sick and I had to take pills. But I don't got it in me no more to fight. I just want to know who my mother was.

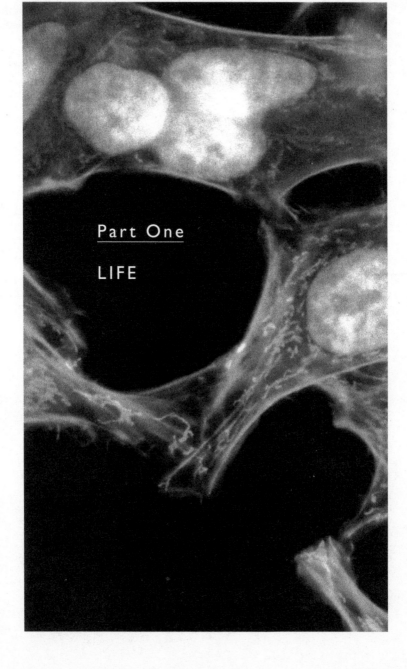

Part One

LIFE

1

The Exam

On January 29, 1951, David Lacks sat behind the wheel of his old Buick, watching the rain fall. He was parked under a towering oak tree outside Johns Hopkins Hospital with three of his children—two still in diapers—waiting for their mother, Henrietta. A few minutes earlier she'd jumped out of the car, pulled her jacket over her head, and scurried into the hospital, past the "colored" bathroom, the only one she was allowed to use. In the next building, under an elegant domed copper roof, a ten-and-a-half-foot marble statue of Jesus stood, arms spread wide, holding court over what was once the main entrance of Hopkins. No one in Henrietta's family ever saw a Hopkins doctor without visiting the Jesus statue, laying flowers at his feet, saying a prayer, and rubbing his big toe for good luck. But that day Henrietta didn't stop.

She went straight to the waiting room of the gynecology clinic, a wide-open space, empty but for rows of long straight-backed benches that looked like church pews.

"I got a knot on my womb," she told the receptionist. "The doctor need to have a look."

For more than a year Henrietta had been telling her closest girl-friends something didn't feel right. One night after dinner, she sat on her bed with her cousins Margaret and Sadie and told them, "I got a knot inside me."

"A what?" Sadie asked.

"A knot," she said. "It hurt somethin awful—when that man want to get with me, Sweet Jesus aren't them but some pains."

When sex first started hurting, she thought it had something to do with baby Deborah, who she'd just given birth to a few weeks earlier, or the bad blood David sometimes brought home after nights with other women—the kind doctors treated with shots of penicillin and heavy metals.

Henrietta grabbed her cousins' hands one at a time and guided them to her belly, just as she'd done when Deborah started kicking.

"You feel anything?"

The cousins pressed their fingers into her stomach again and again.

"I don't know," Sadie said. "Maybe you're pregnant outside your womb—you know that *can* happen."

"I'm no kind of pregnant," Henrietta said. "It's a knot."

"Hennie, you gotta check that out. What if it's somethin bad?"

But Henrietta didn't go to the doctor, and the cousins didn't tell anyone what she'd said in the bedroom. In those days, people didn't talk about things like cancer, but Sadie always figured Henrietta kept it secret because she was afraid a doctor would take her womb and make her stop having children.

About a week after telling her cousins she thought something was wrong, at the age of twenty-nine, Henrietta turned up pregnant with Joe, her fifth child. Sadie and Margaret told Henrietta that the pain probably had something to do with a baby after all. But Henrietta still said no.

"It was there before the baby," she told them. "It's somethin else."

They all stopped talking about the knot, and no one told Henrietta's husband anything about it. Then, four and a half months after baby Joseph was born, Henrietta went to the bathroom and found blood spotting her underwear when it wasn't her time of the month.

She filled her bathtub, lowered herself into the warm water, and spread her legs. With the door closed to her children, husband, and cousins, Henrietta slid a finger inside herself and rubbed it across her cervix until she found what she somehow knew she'd find: a hard lump, deep inside, as though someone had lodged a marble just to the left of the opening to her womb.

Henrietta climbed out of the bathtub, dried herself off, and dressed. Then she told her husband, "You better take me to the doctor. I'm bleedin and it ain't my time."

Her local doctor took one look inside her, saw the lump, and figured it was a sore from syphilis. But the lump tested negative for syphilis, so he told Henrietta she'd better go to the Johns Hopkins gynecology clinic.

Hopkins was one of the top hospitals in the country. It was built in 1889 as a charity hospital for the sick and poor, and it covered more than a dozen acres where a cemetery and insane asylum once sat in East Baltimore. The public wards at Hopkins were filled with patients, most of them black and unable to pay their medical bills. David drove Henrietta nearly twenty miles to get there, not because they preferred it, but because it was the only major hospital for miles that treated black patients. This was the era of Jim Crow—when black people showed up at white-only hospitals, the staff was likely to send them away, even if it meant they might die in the parking lot. Even Hopkins, which did treat black patients, segregated them in colored wards, and had colored-only fountains.

So when the nurse called Henrietta from the waiting room, she led her through a single door to a colored-only exam room—one in a long row of rooms divided by clear glass walls that let nurses see from one to the next. Henrietta undressed, wrapped herself in a starched white hospital gown, and lay down on a wooden exam table, waiting for Howard Jones, the gynecologist on duty. Jones was thin and graying, his deep voice softened by a faint Southern accent. When he walked into the room, Henrietta told him about the lump. Before examining her, he flipped through her chart—a quick sketch of her life, and a litany of untreated conditions:

> Sixth or seventh grade education; housewife and mother of five. Breathing difficult since childhood due to recurrent throat infections and deviated septum in patient's nose. Physician recommended surgical repair. Patient declined. Patient had one toothache for nearly five years; tooth eventually extracted with several others. Only anxiety is oldest daughter who is epileptic and can't talk. Happy household. Very occasional drinker. Has not traveled. Well nourished, cooperative. Patient was one of ten siblings. One died of car accident, one from rheumatic heart, one was poisoned. Unexplained vaginal bleeding and blood in urine during last two pregnancies; physician recommended sickle cell test. Patient declined. Been with husband since age 15 and has no liking for sexual intercourse. Patient has asymptomatic neurosyphilis but cancelled syphilis treatments, said she felt fine. Two months prior to current visit, after delivery of fifth child, patient had significant blood in urine. Tests showed areas of increased cellular activity in the cervix. Physician recommended diagnostics and referred to specialist for ruling out infection or cancer. Patient canceled appointment. One month prior to current visit, patient tested positive for gonorrhea. Patient recalled to clinic for treatment. No response.

It was no surprise that she hadn't come back all those times for follow-up. For Henrietta, walking into Hopkins was like entering a foreign country where she didn't speak the language. She knew about harvesting tobacco and butchering a pig, but she'd never heard the words *cervix* or *biopsy*. She didn't read or write much, and she hadn't studied science in school. She, like most black patients, only went to Hopkins when she thought she had no choice.

Jones listened as Henrietta told him about the pain, the blood. "She says that she knew there was something wrong with the neck of her womb," he wrote later. "When asked why she knew it, she said that she felt as if there were a lump there. I do not quite know what she means by this, unless she actually palpated this area."

Henrietta lay back on the table, feet pressed hard in stirrups as she stared at the ceiling. And sure enough, Jones found a lump exactly where she'd said he would. He described it as an eroded, hard mass about the size of a nickel. If her cervix was a clock's face, the lump was at four o'clock. He'd seen easily a thousand cervical cancer lesions, but never anything like this: shiny and purple (like "grape Jello," he wrote later), and so delicate it bled at the slightest touch. Jones cut a small sample and sent it to the pathology lab down the hall for a diagnosis. Then he told Henrietta to go home.

Soon after, Jones sat down and dictated notes about Henrietta and her diagnosis: "Her history is interesting in that she had a term delivery here at this hospital, September 19, 1950," he said. "No note is made in the history at that time, or at the six weeks' return visit that there is any abnormality of the cervix."

Yet here she was, three months later, with a full-fledged tumor. Either her doctors had missed it during her last exams—which seemed impossible—or it had grown at a terrifying rate.

2

Clover

Henrietta Lacks was born Loretta Pleasant in Roanoke, Virginia, on August 1, 1920. No one knows how she became Henrietta. A midwife named Fannie delivered her into a small shack on a dead-end road overlooking a train depot, where hundreds of freight cars came and went each day. Henrietta shared that house with her parents and eight older siblings until 1924, when her mother, Eliza Lacks Pleasant, died giving birth to her tenth child.

Henrietta's father, Johnny Pleasant, was a squat man who hobbled around on a cane he often hit people with. Family lore has it that he killed his own brother for trying to get fresh with Eliza. Johnny didn't have the patience for raising children, so when Eliza died, he took them all back to Clover, Virginia, where his family still farmed the tobacco fields their ancestors had worked as slaves. No one in Clover could take all ten children, so relatives divided them up—one with this cousin, one with that aunt. Henrietta ended up with her grandfather, Tommy Lacks.

Tommy lived in what everyone called the home-house—a four-room log cabin that once served as slave quarters, with plank floors, gas lanterns, and water Henrietta hauled up a long hill from the creek. The home-house stood on a hillside where wind whipped through cracks in the walls. The air inside stayed so cool that when relatives died, the family kept their corpses in the front hallway for days so people could visit and pay respects. Then they buried them in the cemetery out back.

Henrietta's grandfather was already raising another grandchild that one of his daughters had left behind after delivering him on the home-house floor. That child's name was David Lacks, but everyone called him Day, because in the Lacks country drawl, house sounds like *hyse,* and David sounds like *Day.*

Young Day was what the Lacks family called a sneak baby: a man named Johnny Coleman had passed through town; nine months later Day arrived. A twelve-year-old cousin and midwife named Munchie delivered him, blue as a stormy sky and not breathing. A white doctor came to the home-house with his derby and walking stick, wrote "stillborn" on Day's birth certificate, then drove his horse-drawn buggy back to town, leaving a cloud of red dust behind.

Munchie prayed as he rode away, *Lord, I know you didn't mean to take this baby.* She washed Day in a tub of warm water, then put him on a white sheet where she rubbed and patted his chest until he gasped for breath and his blue skin warmed to soft brown.

By the time Johnny Pleasant shipped Henrietta off to live with Grandpa Tommy, she was four and Day was almost nine. No one could have guessed she'd spend the rest of her life with Day—first as a cousin growing up in their grandfather's home, then as his wife.

As children, Henrietta and Day awoke each morning at four o'clock to milk the cows and feed the chickens, hogs, and horses. They tended a garden filled with corn, peanuts, and greens, then headed to the tobacco fields with their cousins Cliff, Fred, Sadie, Margaret, and a horde of others. They spent much of their young lives stooped in those fields, planting tobacco behind mule-drawn plows. Each harvest they pulled the wide leaves from their stalks and tied them into small bundles—their fingers raw and sticky with nicotine resin—then climbed the rafters of their grandfather's tobacco barn to hang bundle after bundle for curing. Each summer day they prayed for a storm to cool their skin from the burning sun. When they got one, they'd scream and run through fields, snatching armfuls of ripe fruit and walnuts that the winds blew from the trees.

Like most young Lackses, Day didn't finish school: he stopped in the fourth grade because the family needed him to work the fields. But Henrietta stayed until the sixth grade. During the school year, after taking care of the garden and livestock each morning, she'd walk two miles—past the white school where children threw rocks and taunted her—to the colored school, a three-room wooden farmhouse hidden under tall shade trees, with a yard out front where Mrs. Coleman made the boys and girls play on separate sides. When school let out each day, and any time it wasn't in session, Henrietta was in the fields with Day and the cousins.

If the weather was nice, when they finished working, the cousins ran straight to the swimming hole they made each year by damming the creek behind the house with rocks, sticks, bags of sand, and anything else they could sink. They threw rocks to scare away the poisonous cottonmouth snakes, then dropped into the water from tree branches or dove from muddy banks.

At nightfall they built fires with pieces of old shoes to keep the mosquitoes away, and watched the stars from beneath the big oak tree where they'd hung a rope to swing from. They played tag, ring-around-the-rosy, and hopscotch, and danced around the field singing until Grandpa Tommy yelled for everyone to go to bed.

Each night, piles of cousins packed into the crawl space above a little wooden kitchen house just a few feet from the home-house. They lay one next to the other—telling stories about the headless tobacco farmer who roamed the streets at night, or the man with no eyes who lived by the creek—then slept until their grandmother Chloe fired up the woodstove below and woke them to the smell of fresh biscuits.

One evening each month during harvest season, Grandpa Tommy hitched the horses after supper and readied them to ride into the town of South Boston—home of the nation's second-largest tobacco market, with tobacco parades, a Miss Tobacco pageant, and a port where boats collected the dried leaves for people around the world to smoke.

Before leaving home, Tommy would call for the young cousins, who'd nestle into the flat wagon on a bed of tobacco leaves, then fight sleep as long as they could before giving in to the rhythm of the horses. Like farmers from all over Virginia, Tommy Lacks and the grandchildren rode through the night to bring their crops to South Boston, where they'd line up at dawn—one wagon behind the next—waiting for the enormous green wooden gates of the auction warehouse to open.

When they arrived, Henrietta and the cousins would help unhitch the horses and fill their troughs with grain, then unload the family's tobacco onto the wood-plank floor of the warehouse. The auctioneer rattled off numbers that echoed through the huge open room, its ceiling nearly thirty feet high and covered with skylights blackened by years of dirt. As Tommy Lacks stood by his crop praying for a good price, Henrietta and the cousins ran around the tobacco piles, talking in a fast gibberish to sound like the auctioneer. At night they'd help Tommy haul any unsold tobacco down to the basement, where he'd turn the leaves into a bed for the children. White farmers slept upstairs in lofts and private rooms; black farmers slept in the dark underbelly of the warehouse with the horses, mules, and dogs, on a dusty dirt floor lined with rows of wooden stalls for livestock, and mountains of empty liquor bottles piled almost to the ceiling.

Night at the warehouse was a time of booze, gambling, prostitution, and occasional murders as farmers burned through their season's earnings. From their bed of leaves, the Lacks children would stare at ceiling beams the size of trees as they drifted off to the sound of laughter and clanking bottles, and the smell of dried tobacco.

In the morning they'd pile into the wagon with their unsold harvest and set out on the long journey home. Any cousins who'd stayed behind in Clover knew a wagon ride into South Boston meant treats for everyone—a hunk of cheese, maybe, or a slab of bologna—

so they waited for hours on Main Street to follow the wagon to the home-house.

Clover's wide, dusty Main Street was full of Model As, and wagons pulled by mules and horses. Old Man Snow had the first tractor in town, and he drove it to the store like it was a car—newspaper tucked under his arm, his hounds Cadillac and Dan baying beside him. Main Street had a movie theater, bank, jewelry store, doctor's office, hardware store, and several churches. When the weather was good, white men with suspenders, top hats, and long cigars—everyone from mayor to doctor to undertaker—stood along Main Street sipping whiskey from juice bottles, talking, or playing checkers on the wooden barrel in front of the pharmacy. Their wives gossiped at the general store as their babies slept in a row on the counter, heads resting on long bolts of fabric.

Henrietta and her cousins would hire themselves out to those white folks, pulling their tobacco for ten cents so they'd have money to see their favorite Buck Jones cowboy movies. The theater owner showed silent black-and-white films, and his wife played along on the piano. She knew only one song, so she played happy carnival-style music for every scene, even when characters were getting shot and dying. The Lacks children sat up in the colored section next to the projector, which clicked like a metronome through the whole movie.

As Henrietta and Day grew older, they traded ring-around-the-rosy for horse races along the dirt road that ran the length of what used to be the Lacks tobacco plantation, but was now simply called Lacks Town. The boys always fought over who got to ride Charlie Horse, Grandpa Tommy's tall bay, which could outrun any other horse in Clover. Henrietta and the other girls watched from the hillside or the backs of straw-filled wagons, hopping up and down, clapping and screaming as the boys streaked by on horseback.

Henrietta often yelled for Day, but sometimes she cheered for another cousin, Crazy Joe Grinnan. Crazy Joe was what their cousin

Cliff called "an over average man"—tall, husky, and strong, with dark skin, a sharp nose, and so much thick black hair covering his head, arms, back, and neck that he had to shave his whole body in the summer to keep from burning up. They called him Crazy Joe because he was so in love with Henrietta, he'd do anything to get her attention. She was the prettiest girl in Lacks Town, with her beautiful smile and walnut eyes.

The first time Crazy Joe tried to kill himself over Henrietta, he ran circles around her in the middle of winter while she was on her way home from school. He begged her for a date, saying, "Hennie, come on . . . just give me a chance." When she laughed and said no, Crazy Joe ran and jumped straight through the ice of a frozen pond and refused to come out until she agreed to go out with him.

All the cousins teased Joe, saying, "Maybe he thought that ice water might'a cool him off, but he so hot for her, that water nearly started boiling!" Henrietta's cousin Sadie, who was Crazy Joe's sister, yelled at him, "Man you so much in love with a girl, you gonna die for her? That ain't right."

No one knew what happened between Henrietta and Crazy Joe, except that there were some dates and some kisses. But Henrietta and Day had been sharing a bedroom since she was four, so what happened next didn't surprise anyone: they started having children together. Their son Lawrence was born just months after Henrietta's fourteenth birthday; his sister Lucile Elsie Pleasant came along four years later. They were both born on the floor of the home-house like their father, grandmother, and grandfather before them.

People wouldn't use words like *epilepsy, mental retardation,* or *neurosyphilis* to describe Elsie's condition until years later. To the folks in Lacks Town, she was just simple. Touched. She came into the world so fast, Day hadn't even gotten back with the midwife when Elsie shot right out and hit her head on the floor. Everyone would say maybe that was what left her mind like an infant's.

The old dusty record books from Henrietta's church are filled with the names of women cast from the congregation for bearing children out of wedlock, but for some reason Henrietta never was, even as rumors floated around Lacks Town that maybe Crazy Joe had fathered one of her children.

When Crazy Joe found out Henrietta was going to marry Day, he stabbed himself in the chest with an old dull pocketknife. His father found him lying drunk in their yard, shirt soaked with blood. He tried to stop the bleeding, but Joe fought him—thrashing and punching—which just made him bleed more. Eventually Joe's father wrestled him into the car, tied him tight to the door, and drove to the doctor. When Joe got home all bandaged up, Sadie just kept saying, "All that to stop Hennie from marrying Day?" But Crazy Joe wasn't the only one trying to stop the marriage.

Henrietta's sister Gladys was always saying Henrietta could do better. When most Lackses talked about Henrietta and Day and their early life in Clover, it sounded as idyllic as a fairy tale. But not Gladys. No one knew why she was so against the marriage. Some folks said Gladys was just jealous because Henrietta was prettier. But Gladys always insisted Day would be a no-good husband.

Henrietta and Day married alone at their preacher's house on April 10, 1941. She was twenty; he was twenty-five. They didn't go on a honeymoon because there was too much work to do, and no money for travel. By winter, the United States was at war and tobacco companies were supplying free cigarettes to soldiers, so the market was booming. But as large farms flourished, the small ones struggled. Henrietta and Day were lucky if they sold enough tobacco each season to feed the family and plant the next crop.

So after their wedding, Day went back to gripping the splintered ends of his old wooden plow as Henrietta followed close behind, pushing a homemade wheelbarrow and dropping tobacco seedlings into holes in the freshly turned red dirt.

Then one afternoon at the end of 1941, their cousin Fred Garret came barreling down the dirt road beside their field. He was just back from Baltimore for a visit in his slick '36 Chevy and fancy clothes. Only a year earlier, Fred and his brother Cliff had been tobacco farmers in Clover too. For extra money, they'd opened a "colored" convenience store where most customers paid in IOUs; they also ran an old cinderblock juke joint where Henrietta often danced on the red-dirt floor. Everybody put coins in the jukebox and drank RC Cola, but the profits never amounted to much. So eventually Fred took his last three dollars and twenty-five cents and bought a bus ticket north for a new life. He, like several other cousins, went to work at Bethlehem Steel's Sparrows Point steel mill and live in Turner Station, a small community of black workers on a peninsula in the Patapsco River, about twenty miles from downtown Baltimore.

In the late 1800s, when Sparrows Point first opened, Turner Station was mostly swamps, farmland, and a few shanties connected with wooden boards for walkways. When demand for steel increased during World War I, streams of white workers moved into the nearby town of Dundalk, and Bethlehem Steel's housing barracks for black workers quickly overflowed, pushing them into Turner Station. By the early years of World War II, Turner Station had a few paved roads, a doctor, a general store, and an ice man. But its residents were still fighting for water, sewage lines, and schools.

Then, in December 1941, Japan bombed Pearl Harbor, and it was like Turner Station had won the lottery: the demand for steel skyrocketed, as did the need for workers. The government poured money into Turner Station, which began filling with one- and two-story housing projects, many of them pressed side by side and back-to-back, some with four to five hundred units. Most were brick, others covered with asbestos shingles. Some had yards, some didn't. From most of them you could see the flames dancing above Sparrows Point's furnaces and the eerie red smoke pouring from its smokestacks.

Sparrows Point was rapidly becoming the largest steel plant in the world. It produced concrete-reinforcing bars, barbed wire, nails, and steel for cars, refrigerators, and military ships. It would burn more than six million tons of coal each year to make up to eight million tons of steel and employ more than 30,000 workers. Bethlehem Steel was a gold mine in a time flush with poverty, especially for black families from the South. Word spread from Maryland to the farms of Virginia and the Carolinas, and as part of what would become known as the Great Migration, black families flocked from the South to Turner Station—the Promised Land.

The work was tough, especially for black men, who got the jobs white men wouldn't touch. Like Fred, black workers usually started in the bowels of partially built tankers in the shipyard, collecting bolts, rivets, and nuts as they fell from the hands of men drilling and welding thirty or forty feet up. Eventually black workers moved up to the boiler room, where they shoveled coal into a blazing furnace. They spent their days breathing in toxic coal dust and asbestos, which they brought home to their wives and daughters, who inhaled it while shaking the men's clothes out for the wash. The black workers at Sparrows Point made about eighty cents an hour at most, usually less. White workers got higher wages, but Fred didn't complain: eighty cents an hour was more than most Lackses had ever seen.

Fred had made it. Now he'd come back to Clover to convince Henrietta and Day that they should do the same. The morning after he came barreling into town, Fred bought Day a bus ticket to Baltimore. They agreed Henrietta would stay behind to care for the children and the tobacco until Day made enough for a house of their own in Baltimore, and three tickets north. A few months later, Fred got a draft notice shipping him overseas. Before he left, Fred gave Day all the money he'd saved, saying it was time to get Henrietta and the children to Turner Station.

Soon, with a child on each side, Henrietta boarded a coal-fueled train from the small wooden depot at the end of Clover's Main Street. She left the tobacco fields of her youth and the hundred-year-old oak tree that shaded her from the sun on so many hot afternoons. At the age of twenty-one, Henrietta stared through the train window at rolling hills and wide-open bodies of water for the first time, heading toward a new life.

3

Diagnosis and Treatment

After her visit to Hopkins, Henrietta went about life as usual, cleaning and cooking for Day, their children, and the many cousins who stopped by. Then, a few days later, Jones got her biopsy results from the pathology lab: "Epidermoid carcinoma of the cervix, Stage I."

All cancers originate from a single cell gone wrong and are categorized based on the type of cell they start from. Most cervical cancers are carcinomas, which grow from the epithelial cells that cover the cervix and protect its surface. By chance, when Henrietta showed up at Hopkins complaining of abnormal bleeding, Jones and his boss, Richard Wesley TeLinde, were involved in a heated nationwide debate over what qualified as cervical cancer, and how best to treat it.

TeLinde, one of the top cervical cancer experts in the country, was a dapper and serious fifty-six-year-old surgeon who walked with an extreme limp from an ice-skating accident more than a decade earlier. Everyone at Hopkins called him Uncle Dick. He'd pioneered the use of estrogen for treating symptoms of menopause and made important early discoveries about endometriosis. He'd also written one of the most famous clinical gynecology textbooks, which is still widely used sixty years and ten editions after he first wrote it. His reputation was international: when the king of Morocco's wife fell ill, he insisted only TeLinde could operate on her. By 1951, when Henrietta arrived at Hopkins, TeLinde had developed a theory about cervical cancer that, if correct, could save the lives of millions of women. But few in the field believed him.

Cervical carcinomas are divided into two types: invasive carcinomas, which have penetrated the surface of the cervix, and noninvasive carcinomas, which haven't. The noninvasive type is sometimes called "sugar-icing carcinoma," because it grows in a smooth layered sheet across the surface of the cervix, but its official name is *carcinoma in situ,* which derives from the Latin for "cancer in its original place."

In 1951, most doctors in the field believed that invasive carcinoma was deadly, and carcinoma in situ wasn't. So they treated the invasive type aggressively but generally didn't worry about carcinoma in situ because they thought it couldn't spread. TeLinde disagreed—he believed carcinoma in situ was simply an early stage of invasive carcinoma that, if left untreated, eventually became deadly. So he treated it aggressively, often removing the cervix, uterus, and most of the vagina. He argued that this would drastically reduce cervical cancer deaths, but his critics called it extreme and unnecessary.

Diagnosing carcinoma in situ had only been possible since 1941, when George Papanicolaou, a Greek researcher, published a paper describing a test he'd developed, now called the Pap smear. It involved scraping cells from the cervix with a curved glass pipette and examining them under a microscope for precancerous changes that TeLinde and a few others had identified years earlier. This was a tremendous advance, because those precancerous cells weren't detectable otherwise: they caused no physical symptoms and weren't palpable or visible to the naked eye. By the time a woman began showing symptoms, there was little hope of a cure. But with the Pap smear, doctors could detect precancerous cells and perform a hysterectomy, and cervical cancer would be almost entirely preventable.

At that point, more than 15,000 women were dying each year from cervical cancer. The Pap smear had the potential to decrease that death rate by 70 percent or more, but there were two things standing in its way: first, many women—like Henrietta—simply didn't get the test; and, second, even when they did, few doctors knew how to interpret the results accurately, because they didn't know what the various stages of cervical cancer looked like under a microscope. Some mistook cervical infections for cancer and removed a woman's entire reproductive tract when all she needed was antibiotics. Others mistook malignant changes for infection, sending women home with antibiotics only to have them return later, dying from metastasized cancer. And even when doctors correctly diagnosed precancerous changes, they often didn't know how those changes should be treated.

TeLinde set out to minimize what he called "unjustifiable hysterectomies" by documenting what *wasn't* cervical cancer and by urging surgeons to verify smear results with biopsies before operating. He also hoped to prove that women with carcinoma in situ needed aggressive treatment, so their cancer didn't become invasive.

Not long before Henrietta's first exam, TeLinde presented his argument about carcinoma in situ to a major meeting of pathologists in Washington, D.C., and the audience heckled him off the stage. So he went back to Hopkins and planned a study that would prove them wrong: he and his staff would review all medical records and biopsies from patients who'd been diagnosed with invasive cervical cancer at Hopkins in the past decade, to see how many initially had carcinoma in situ.

Like many doctors of his era, TeLinde often used patients from the public wards for research, usually without their knowledge. Many scientists believed that since patients were treated for free in the public wards, it was fair to use them as research subjects as a form of payment. And as Howard Jones once wrote, "Hopkins, with its large indigent black population, had no dearth of clinical material."

In this particular study—the largest ever done on the relationship between the two cervical cancers—Jones and TeLinde found that 62 percent of women with invasive cancer who'd had earlier biopsies first had carcinoma in situ. In addition to that study, TeLinde thought, if he could find a way to grow living samples from normal cervical tissue and both types of cancerous tissue—something never done before—he could compare all three. If he could prove that carcinoma in situ and invasive carcinoma looked and behaved similarly in the laboratory, he could end the debate, showing that he'd been right all along, and doctors who ignored him were killing their patients. So he called George Gey (pronounced *Guy*), head of tissue culture research at Hopkins.

Gey and his wife, Margaret, had spent the last three decades working to grow malignant cells outside the body, hoping to use them to find cancer's cause and cure. But most cells died quickly, and the few that survived hardly grew at all. The Geys were determined to grow the first *immortal* human cells: a continuously dividing line of cells all descended from one original sample, cells that would constantly replenish themselves and never die. Eight years earlier—in 1943—a group of researchers at the National Institutes of Health had proven such a thing was possible using mouse cells. The Geys wanted to grow the human equivalent—they didn't care what kind of tissue they used, as long as it came from a person.

Gey took any cells he could get his hands on—he called himself "the world's most famous vulture, feeding on human specimens almost constantly." So when TeLinde offered him a supply of cervical cancer tissue in exchange for trying to grow some cells, Gey didn't hesitate. And TeLinde began collecting samples from any woman who happened to walk into Hopkins with cervical cancer. Including Henrietta.

On February 5, 1951, after Jones got Henrietta's biopsy report back from the lab, he called and told her it was malignant. Henrietta didn't tell anyone what Jones said, and no one asked. She simply went on with her day as if nothing had happened, which was just like her—no sense upsetting anyone over something she could deal with herself.

That night Henrietta told her husband, "Day, I need to go back to the doctor tomorrow. He wants to do some tests, give me some medicine." The next morning she climbed from the Buick outside Hopkins again, telling Day and the children not to worry.

"Ain't nothin serious wrong," she said. "Doctor's gonna fix me right up."

Henrietta went straight to the admissions desk and told the receptionist she was there for her treatment. Then she signed a form with the words OPERATION PERMIT at the top of the page. It said:

> I hereby give consent to the staff of The Johns Hopkins Hospital to perform any operative procedures and under any anaesthetic either local or general that they may deem necessary in the proper surgical care and treatment of: _____.

Henrietta printed her name in the blank space. A witness with illegible handwriting signed a line at the bottom of the form, and Henrietta signed another.

Then she followed a nurse down a long hallway into the ward for colored women, where Howard Jones and several other white physicians ran more tests than she'd had in her entire life. They checked her urine, her blood, her lungs. They stuck tubes in her bladder and nose.

On her second night at the hospital, the nurse on duty fed Henrietta an early dinner so her stomach would be empty the next morning, when a doctor put her under anesthetic for her first cancer treatment. Henrietta's tumor was the invasive type, and like hospitals nationwide,

Hopkins treated all invasive cervical carcinomas with radium, a white radioactive metal that glows an eerie blue.

When radium was first discovered in the late 1800s, headlines nationwide hailed it as "a substitute for gas, electricity, and a positive cure for every disease." Watchmakers added it to paint to make watch dials glow, and doctors administered it in powdered form to treat everything from seasickness to ear infections. But radium destroys any cells it encounters, and patients who'd taken it for trivial problems began dying. Radium causes mutations that can turn into cancer, and at high doses it can burn the skin off a person's body. But it also kills cancer cells.

Hopkins had been using radium to treat cervical cancer since the early 1900s, when a surgeon named Howard Kelly visited Marie and Pierre Curie, the couple in France who'd discovered radium and its ability to destroy cancer cells. Without realizing the danger of contact with radium, Kelly brought some back to the United States in his pockets and regularly traveled the world collecting more. By the 1940s, several studies—one of them conducted by Howard Jones, Henrietta's physician—showed that radium was safer and more effective than surgery for treating invasive cervical cancer.

The morning of Henrietta's first treatment, a taxi driver picked up a doctor's bag filled with thin glass tubes of radium from a clinic across town. The tubes were tucked into individual slots inside small canvas pouches hand-sewn by a local Baltimore woman. The pouches were called Brack plaques, after the Hopkins doctor who invented them and oversaw Henrietta's radium treatment. He would later die of cancer, most likely caused by his regular exposure to radium, as would a resident who traveled with Kelly and also transported radium in his pockets.

One nurse placed the Brack plaques on a stainless-steel tray. Another wheeled Henrietta into the small colored-only operating room on the second floor, with stainless-steel tables, huge glaring lights, and an all-white medical staff dressed in white gowns, hats, masks, and gloves.

With Henrietta unconscious on the operating table in the center of the room, her feet in stirrups, the surgeon on duty, Dr. Lawrence Wharton Jr., sat on a stool between her legs. He peered inside Henrietta, dilated her cervix, and prepared to treat her tumor. But first—though no one had told Henrietta that TeLinde was collecting samples or asked if she wanted to be a donor—Wharton picked up a sharp knife and shaved two dime-sized pieces of tissue from Henrietta's cervix: one from her tumor, and one from the healthy cervical tissue nearby. Then he placed the samples in a glass dish.

Wharton slipped a tube filled with radium inside Henrietta's cervix, and sewed it in place. He sewed a plaque filled with radium to the outer surface of her cervix and packed another plaque against it. He slid several rolls of gauze inside her vagina to help keep the radium in place, then threaded a catheter into her bladder so she could urinate without disturbing the treatment.

When Wharton finished, a nurse wheeled Henrietta back into the ward, and Wharton wrote in her chart, "The patient tolerated the procedure well and left the operating room in good condition." On a separate page he wrote, "Henrietta Lacks . . . Biopsy of cervical tissue . . . Tissue given to Dr. George Gey."

A resident took the dish with the samples to Gey's lab, as he'd done many times before. Gey still got excited at moments like this, but everyone else in his lab saw Henrietta's sample as something tedious—the latest of what felt like countless samples that scientists and lab technicians had been trying and failing to grow for years. They were sure Henrietta's cells would die just like all the others.

4

The Birth
of HeLa

Gey's twenty-one-year-old assistant, Mary Kubicek, sat eating a tuna-salad sandwich at a long stone culture bench that doubled as a break table. She and Margaret and the other women in the Gey lab spent countless hours there, all in nearly identical cat-eye-glasses with fat dark frames and thick lenses, their hair pulled back in tight buns.

At first glance, the room could have been an industrial kitchen. There were gallon-sized tin coffee cans full of utensils and glassware; powdered creamer, sugar, spoons, and soda bottles on the table; huge metal freezers lining one wall; and deep sinks Gey made by hand using stones he collected from a nearby quarry. But the teapot sat next to a Bunsen burner, and the freezers were filled with blood, placentas, tumor samples, and dead mice (plus at least one duck Gey kept frozen in the lab for more than twenty years after a hunting trip, since it wouldn't fit in his freezer at home). Gey had lined one wall with cages full of squealing rabbits, rats, and guinea pigs; on one side of the table where Mary sat eating her lunch, he'd built shelves holding cages full of mice, their bodies filled with tumors. Mary always stared at them while she ate, just as she was doing when Gey walked into the lab carrying the pieces of Henrietta's cervix.

"I'm putting a new sample in your cubicle," he told her.

Mary pretended not to notice. *Not again,* she thought, and kept eating her sandwich. *It can wait till I'm done.*

Mary knew she shouldn't wait—every moment those cells sat in the dish made it more likely they'd die. But she was tired of cell culture, tired of meticulously cutting away dead tissue like gristle from a steak, tired of having cells die after hours of work.

Why bother? she thought.

Gey hired Mary for her hands. She was fresh out of college with a physiology degree when her adviser sent her for an interview. Gey asked Mary to pick up a pen from the table and write a few sentences. Now pick up that knife, he said. Cut this piece of paper. Twirl this pipette.

Mary didn't realize until months later that he'd been studying her hands, checking their dexterity and strength to see how they'd stand up to hours of delicate cutting, scraping, tweezing, and pipetting.

By the time Henrietta walked into Hopkins, Mary was handling most of the tissue samples that came through the door, and so far all samples from TeLinde's patients had died.

At that point, there were many obstacles to growing cells successfully. For starters, no one knew exactly what nutrients they needed to survive, or how best to supply them. Many researchers, including the Geys, had been trying for years to develop the perfect culture medium—the liquid used for feeding cells. The recipes for Gey Culture Medium evolved constantly as George and Margaret added and removed ingredients, searching for the perfect balance. But they all sounded like witches' brews: the plasma of chickens, purée of calf fetuses, special salts, and blood from human umbilical cords. George had rigged a bell and cable from the window of his lab across a courtyard to the Hopkins maternity ward, so nurses could ring anytime a baby was born, and Margaret or Mary would run over and collect umbilical cord blood.

The other ingredients weren't so easy to come by: George visited local slaughterhouses at least once a week to collect cow fetuses and chicken blood. He'd drive there in his rusted-out old Chevy, its left fender flapping against the pavement, shooting sparks. Well before dawn, in a rundown wooden building with a sawdust floor and wide gaps in the walls, Gey would grab a screaming chicken by the legs, yank it upside down from its cage, and wrestle it to its back on a butcher block. He'd hold its feet in one hand and pin its neck motionless to the wood with his elbow. With his free hand, he'd squirt the bird's chest with alcohol, and plunge a syringe needle into the chicken's heart to draw blood. Then he'd stand the bird upright, saying, "Sorry, old fella," and put it back in its cage. Every once in a while, when a chicken dropped dead from the stress, George took it home so Margaret could fry it for dinner.

Like many procedures in their lab, the Gey Chicken Bleeding Technique was Margaret's creation. She worked out the method step-by-step, taught it to George, and wrote detailed instructions for the many other researchers who wanted to learn it.

Finding the perfect medium was an ongoing experiment, but the biggest problem facing cell culture was contamination. Bacteria and a host of other microorganisms could find their way into cultures from people's unwashed hands, their breath, and dust particles floating through the air, and destroy them. But Margaret had been trained as a surgical nurse, which meant sterility was her specialty—it was key to preventing deadly infections in patients in the operating room. Many would later say that Margaret's surgical training was the only reason the Gey lab was able to grow cells at all. Most culturists, like George, were biologists; they knew nothing about preventing contamination.

Margaret taught George everything he knew about keeping cultures sterile, and she did the same with every technician, grad student, and scientist who came to work or study in the lab. She hired a local woman named Minnie whose sole job was washing the laboratory glassware using the only product Margaret would allow: Gold Dust Twins soap. Margaret was so serious about that soap, when she heard a rumor that the company might go out of business, she bought an entire boxcar full of it.

Margaret patrolled the lab, arms crossed, and leaned over Minnie's shoulder as she worked, towering nearly a foot above her. If Margaret ever smiled, no one could have seen it through her ever-present surgical mask. She inspected all the glassware for spots or smudges, and when she found them—which was often—she'd scream, "MINNIE!" so loud that Mary cringed.

Mary followed Margaret's sterilizing rules meticulously to avoid her wrath. After finishing her lunch, and before touching Henrietta's sample, Mary covered herself with a clean white gown, surgical cap, and mask, and then walked to her cubicle, one of four airtight rooms George had built by hand in the center of the lab. The cubicles were small, only five feet in any direction, with doors that sealed like a freezer's to prevent contaminated air from getting inside. Mary turned on the sterilizing system and watched from outside as her cubicle filled with hot steam to kill anything that might damage the cells. When the steam cleared, she stepped inside and sealed the door behind her, then hosed the cubicle's cement floor with water and scoured her workbench with alcohol. The air inside was filtered and piped in through a vent on the ceiling. Once she'd sterilized the cubicle, she lit a Bunsen burner and used its flame to sterilize test tubes and a used scalpel blade, since the Gey lab couldn't afford new ones for each sample.

Only then did she pick up the pieces of Henrietta's cervix—forceps in one hand, scalpel in the other—and carefully slice them into one-millimeter squares. She sucked each square into a pipette, and dropped them one at a time onto chicken-blood clots she'd placed at the bottom of dozens of test tubes. She covered each clot with several drops of culture medium, plugged the tubes with rubber stoppers, and labeled each one as she'd labeled most cultures they grew: using the first two letters of the patient's first and last names.

After writing "HeLa," for *Henrietta* and *Lacks,* in big black letters on the side of each tube, Mary carried them to the incubator room that Gey had built just like he'd built everything else in the lab: by hand and mostly from junkyard scraps, a skill he'd learned from a lifetime of making do with nothing.

George Gey was born in 1899 and raised on a Pittsburgh hillside overlooking a steel mill. Soot from the smokestacks made his parents' small white house look like it had been permanently charred by fire and left the afternoon sky dark. His mother worked the garden and fed her family from nothing but the food she raised. As a child, George dug a small coal mine in the hill behind his parents' house. He'd crawl through the damp tunnel each morning with a pick, filling buckets for his family and neighbors so they could keep their houses warm and stoves burning.

Gey paid his way through a biology degree at the University of Pittsburgh by working as a carpenter and mason, and he could make nearly anything for cheap or free. During his second year in medical school, he rigged a microscope with a time-lapse motion picture camera to capture live cells on film. It was a Frankensteinish mishmash of microscope parts, glass, and 16-millimeter camera equipment from who knows where, plus metal scraps, and old motors from Shapiro's junkyard. He built it in a hole he'd blasted in the foundation of Hopkins, right below the morgue, its base entirely underground and surrounded by a thick wall of cork to keep it from jiggling when streetcars passed. At night, a Lithuanian lab assistant slept next to the camera on a cot, listening to its constant tick, making sure it stayed stable through the night, waking every hour to refocus it. With that camera, Gey and his mentor, Warren Lewis, filmed the growth of cells, a process so slow—like the growth of a flower—the naked eye couldn't see it. They played the film at high speed so they could watch cell division on the screen in one smooth motion, like a story unfolding in a flip book.

It took Gey eight years to get through medical school because he kept dropping out to work construction and save for another year's tuition. After he graduated, he and Margaret built their first lab in a janitor's quarters at Hopkins—they spent weeks wiring, painting, plumbing, building counters and cabinets, paying for much of it with their own money.

Margaret was cautious and stable, the backbone of the lab. George was an enormous, mischievous, grown-up kid. At work he was dapper, but at home he lived in flannels, khakis, and suspenders. He moved boulders around his yard on weekends, ate twelve ears of corn in one sitting, and kept barrels full of oysters in his garage so he could shuck and eat them anytime he wanted. He had the body of a retired linebacker, six feet four inches tall and 215 pounds, his back unnaturally stiff and upright from having his spine fused so he'd stop throwing it out. When his basement wine-making factory exploded on a Sunday, sending a flood of sparkling burgundy through his garage and into the street, Gey just washed the wine into a storm drain, waving at his neighbors as they walked to church.

Gey was a reckless visionary—spontaneous, quick to start dozens of projects at once, filling the lab and his basement at home with half-built machines, partial discoveries, and piles of junkyard scraps only he could imagine using in a lab. Whenever an idea hit him, he sat wherever he was—at his desk, kitchen table, a bar, or behind the wheel of his car—gnawing on his ever-present cigar and scribbling diagrams on napkins or the backs of torn-off bottle labels. That's how he came up with the roller-tube culturing technique, his most important invention.

It involved a large wooden roller drum, a cylinder with holes for special test tubes called roller tubes. The drum, which Gey called the "whirligig," turned like a cement mixer twenty-four hours a day, rotating so slowly it made only two full turns an hour, sometimes less. For Gey, the rotation was crucial: he believed that culture medium needed to be in constant motion, like blood and fluids in the body, which flow around cells, transporting waste and nutrients.

When Mary finally finished cutting the samples of Henrietta's

cervix and dropping them in dozens of roller tubes, she walked into
the incubator room, slid the tubes one at a time into the drum, and
turned it on. Then she watched as Gey's machine began churning
slowly.

Henrietta spent the next two days in the hos-
pital, recovering from her first radium treatment. Doctors examined
her inside and out, pressing on her stomach, inserting new catheters
into her bladder, fingers into her vagina and anus, needles into her
veins. They wrote notes in her chart saying, "30 year-old colored
female lying quietly in no evident distress," and "Patient feels quite
well tonight. Morale is good and she is ready to go home."

Before Henrietta left the hospital, a doctor put her feet in the stirrups again and removed the radium. He sent her home with instructions to call the clinic if she had problems, and to come back for a
second dose of radium in two and a half weeks.

Meanwhile, each morning after putting Henrietta's cells in culture,
Mary started her days with the usual sterilization drill. She peered
into the tubes, laughing to herself and thinking, *Nothing's happening.
Big surprise.* Then, two days after Henrietta went home from the hospital, Mary saw what looked like little rings of fried egg white around
the clots at the bottoms of each tube. The cells were growing, but
Mary didn't think much of it—other cells had survived for a while in
the lab.

But Henrietta's cells weren't merely surviving, they were growing
with mythological intensity. By the next morning they'd doubled.
Mary divided the contents of each tube into two, giving them room
to grow, and within twenty-four hours, they'd doubled again. Soon
she was dividing them into four tubes, then six. Henrietta's cells grew
to fill as much space as Mary gave them.

Still, Gey wasn't ready to celebrate. "The cells could die any
minute," he told Mary.

But they didn't. They kept growing like nothing anyone had seen, doubling their numbers every twenty-four hours, stacking hundreds on top of hundreds, accumulating by the millions. "Spreading like crabgrass!" Margaret said. They grew twenty times faster than Henrietta's normal cells, which died only a few days after Mary put them in culture. As long as they had food and warmth, Henrietta's cancer cells seemed unstoppable.

Soon, George told a few of his closest colleagues that he thought his lab might have grown the first immortal human cells.

To which they replied, Can I have some? And George said yes.

5

"Blackness Be Spreadin All Inside"

Henrietta knew nothing about her cells growing in a laboratory. After leaving the hospital, she went back to life as usual. She'd never loved the city, so almost every weekend she took the children back to Clover, where she worked the tobacco fields and spent hours churning butter on the steps of the home-house. Though radium often causes relentless nausea, vomiting, weakness, and anemia, there's no record of Henrietta having any side effects, and no one remembers her complaining of feeling sick.

When she wasn't in Clover, Henrietta spent her time cooking for Day, the children, and whichever cousins happened to be at her house. She made her famous rice pudding and slow-cooked greens, chitlins, and the vats of spaghetti with meatballs she kept going on the stove for whenever cousins dropped by hungry. When Day wasn't working the night shift, he and Henrietta spent evenings at home, playing cards and listening to Bennie Smith play blues guitar on the radio after the kids went to sleep. On the nights Day worked, Henrietta and Sadie would wait until the door slammed, count to one hundred, then jump out of bed, put on their dancing clothes, and sneak out of the house, careful not to wake the children. Once they got outside, they'd wiggle their hips and squeal, scampering down the street to the dance floors at Adams Bar and Twin Pines.

"We used to really swing out heavy," Sadie told me years later. "We couldn't help it. They played music that when you heard it just put your soul into it. We'd two-step across that floor, jiggle to some blues, then somebody maybe put a quarter in there and play a slow music song, and Lord we'd just get out there and shake and turn around and all like that!" She giggled like a young girl. "It was some beautiful times." And they were beautiful women.

Henrietta had walnut eyes, straight white teeth, and full lips. She was a sturdy woman with a square jaw, thick hips, short, muscular legs, and hands rough from tobacco fields and kitchens. She kept her nails short so bread dough wouldn't stick under them when she kneaded it, but she always painted them a deep red to match her toenails.

Henrietta spent hours taking care of those nails, touching up chips and brushing on new coats of polish. She'd sit on her bed, polish in hand, hair high on her head in curlers, wearing the silky slip she loved so much she hand-washed it each night. She never wore pants, and rarely left the house without pulling on a carefully pressed skirt and shirt, sliding her feet into her tiny, open-toed pumps, and pinning her hair up with a little flip at the bottom, "just like it was dancin toward her face," Sadie always said.

"Hennie made life come alive—bein with her was like bein with fun," Sadie told me, staring toward the ceiling as she talked. "Hennie just love peoples. She was a person that could really make the good things come out of you."

But there was one person Henrietta couldn't bring out any good in. Ethel, the wife of their cousin Galen, had recently come to Turner Station from Clover, and she hated Henrietta—her cousins always said it was jealousy.

"I guess I can't say's I blame her," Sadie said. "Galen, that husband of Ethel's, he was likin Hennie more than he like Ethel. Lord, he followed Hennie! Everywhere she go, there go Galen—he tried to stay up at Hennie house all the time when Day gone to work. Lord, Ethel *was* jealous—made her hateful to Hennie somethin fierce. Always seemed like she wanted to hurt Hennie." So Henrietta and Sadie would giggle and slip out the back to another club anytime Ethel showed up.

When they weren't sneaking out, Henrietta, Sadie, and Sadie's sister Margaret spent evenings in Henrietta's living room, playing bingo, yelling, and laughing over a pot of pennies while Henrietta's babies— David Jr., Deborah, and Joe—played with the bingo chips on the carpet beneath the table. Lawrence was nearly sixteen, already out having a life of his own. But one child was missing: Henrietta's oldest daughter, Elsie.

Before Henrietta got sick, she took Elsie down to Clover every time she went. Elsie would sit on the stoop of the home-house, staring into the hills and watching the sunrise as Henrietta worked in the garden. She was beautiful, delicate and feminine like Henrietta, who dressed her in homemade outfits with bows and spent hours braiding her long brown curls. Elsie never talked, she just cawed and chirped like a bird as she waved her hands inches from her face. She had wide chestnut eyes that everyone stared into, trying to understand what went on in that pretty head. But she just stared back, unflinching, her eyes haunted with fear and sadness that only softened when Henrietta rocked her back and forth.

Sometimes Elsie raced through the fields, chasing wild turkeys or grabbing the family mule by the tail and thrashing against him until Lawrence pulled her off. Henrietta's cousin Peter always said God had that child from the moment she was born, because that mule never hurt her. It was so mean it snapped at air like a rabid dog and kicked at the wind, but it seemed to know Elsie was special. Still, as she grew, she fell, she ran into walls and doors, burned herself against the woodstove. Henrietta made Day drive her and Elsie to revival meetings so preachers in tents could lay hands on Elsie to heal her, but it never worked. In Turner Station, sometimes Elsie bolted from the house and ran through the street screaming.

By the time Henrietta got pregnant with baby Joe, Elsie was too big for Henrietta to handle alone, especially with two babies. The doctors said that sending Elsie away was the best thing. So now she was living about an hour and a half south of Baltimore, at Crownsville State Hospital—formerly known as the Hospital for the Negro Insane.

Henrietta's cousins always said a bit of Henrietta died the day they sent Elsie away, that losing her was worse than anything else that happened to her. Now, nearly a year later, Henrietta still had Day or a cousin take her from Turner Station to Crownsville once a week to sit with Elsie, who'd cry and cling to her as they played with each other's hair.

Henrietta had a way with children—they were always good and quiet when she was around. But whenever she left the house, Lawrence stopped being good. If the weather was nice, he'd run to the old pier in Turner Station, where Henrietta had forbidden him to go. The pier had burned down years earlier, leaving tall wooden pilings that Lawrence and his friends liked to dive from. One of Sadie's sons nearly drowned out there from hitting his head on a rock, and Lawrence was always coming home with eye infections that everyone blamed on the water being contaminated by Sparrows Point. Anytime Henrietta got word that Lawrence was at the pier, she'd storm down there, drag him out of the water, and whip him.

"*Ooooh* Lord," Sadie said once, "Hennie went down there with a switch. Yes *Lord*. She pitched a boogie like I never seen." But those were the only times anyone could ever remember seeing Henrietta mad. "She was tough," Sadie said. "Nothin scared Hennie."

For a month and a half, no one in Turner Station knew Henrietta was sick. The cancer was easy to keep secret, because she only had to go back to Hopkins once, for a checkup and a second radium treatment. At that point the doctors liked what they saw: her cervix was a bit red and inflamed from the first treatment, but the tumor was shrinking. Regardless, she had to start X-ray therapy, which meant visiting Hopkins every weekday for a month. For that, she needed help: Henrietta lived twenty minutes from Hopkins, and Day worked nights, so he couldn't take her home after radiation until late. She wanted to walk to her cousin Margaret's house a few blocks from Hopkins and wait there for Day after her treatments. But first she'd have to tell Margaret and Sadie she was sick.

Henrietta told her cousins about the cancer at a carnival that came to Turner Station each year. The three of them climbed onto the Ferris wheel as usual, and she waited till it got so high they could see across Sparrows Point toward the ocean, till the Ferris wheel stopped and they were just kicking their legs back and forth, swinging in the crisp spring air.

"You remember when I said I had a knot inside me?" she asked. They nodded yes. "Well, I got cancer," Henrietta said. "I been havin treatments down at John Hopkins."

"What?!" Sadie said, looking at Henrietta and feeling suddenly dizzy, like she was about to slide off the Ferris wheel seat.

"Nothin serious wrong with me," Henrietta said. "I'm fine."

And at that point it looked like she was right. The tumor had completely vanished from the radium treatments. As far as the doctors could see, Henrietta's cervix was normal again, and they felt no tumors anywhere else. Her doctors were so sure of her recovery that while she was in the hospital for her second radium treatment, they'd performed reconstructive surgery on her nose, fixing the deviated septum that had given her sinus infections and headaches her whole life. It was a new beginning. The radiation treatments were just to make sure there were no cancer cells left anywhere inside her.

But about two weeks after her second radium treatment, Henrietta got her period—the flow was heavy and it didn't stop. She was still bleeding weeks later on March 20, when Day began dropping her off each morning at Hopkins for her radiation treatments. She'd change into a surgical gown, lie on an exam table with an enormous machine mounted on the wall above her, and a doctor would put strips of lead inside her vagina to protect her colon and lower spine from the radiation. On the first day he tattooed two black dots with temporary ink on either side of her abdomen, just over her uterus. They were targets, so he could aim the radiation into the same area each day, but rotate between spots to avoid burning her skin too much in one place.

After each treatment, Henrietta would change back into her clothes and walk the few blocks to Margaret's house, where she'd wait for Day to pick her up around midnight. For the first week or so, she and Margaret would sit on the porch playing cards or bingo, talking about the men, the cousins, and the children. At that point, the radiation seemed like nothing more than an inconvenience. Henrietta's bleeding stopped, and if she felt sick from the treatments, she never mentioned it.

But things weren't all good. Toward the end of her treatments, Henrietta asked her doctor when she'd be better so she could have another child. Until that moment, Henrietta didn't know that the treatments had left her infertile.

Warning patients about fertility loss before cancer treatment was standard practice at Hopkins, and something Howard Jones says he and TeLinde did with every patient. In fact, a year and a half before Henrietta came to Hopkins for treatment, in a paper about hysterectomy, TeLinde wrote:

> The psychic effect of hysterectomy, especially on the young, is considerable, and it should not be done without a thorough understanding on the part of the patient [who is] entitled to a simple explanation of the facts [including] loss of the reproductive function. . . . It is well to present the facts to such an individual and give her ample time to digest them. . . . It is far better for her to make her own adjustment before the operation than to awaken from the anesthetic and find it a *fait accompli.*

In this case, something went wrong: in Henrietta's medical record, one of her doctors wrote, "Told she could not have any more children.

Says if she had been told so before, she would not have gone through with treatment." But by the time she found out, it was too late.

Then, three weeks after starting X-ray therapy, she began burning inside, and her urine came out feeling like broken glass. Day said he'd been having a funny discharge, and that she must have given him that sickness she kept going to Hopkins to treat.

"I would rather imagine that it is the other way around," Jones wrote in Henrietta's chart after examining her. "But at any rate, this patient now has . . . acute Gonorrhea superimposed on radiation reaction."

Soon, however, Day's running around was the least of Henrietta's worries. That short walk to Margaret's started feeling longer and longer, and all Henrietta wanted to do when she got there was sleep. One day she almost collapsed a few blocks from Hopkins, and it took her nearly an hour to make the walk. After that, she started taking cabs.

One afternoon, as Henrietta lay on the couch, she lifted her shirt to show Margaret and Sadie what the treatments had done to her. Sadie gasped: The skin from Henrietta's breasts to her pelvis was charred a deep black from the radiation. The rest of her body was its natural shade—more the color of fawn than coal.

"Hennie," she whispered, "they burnt you black as tar."

Henrietta just nodded and said, "Lord, it just feels like that blackness be spreadin all inside me."

6

"Lady's on
the Phone"

Eleven years after learning about Henrietta in
Defler's classroom—on my twenty-seventh birthday—I stumbled on
a collection of scientific papers from something called "The HeLa
Cancer Control Symposium" at Morehouse School of Medicine in
Atlanta, one of the oldest historically black colleges in the country.
The symposium had been organized in Henrietta's honor by Roland
Pattillo, a professor of gynecology at Morehouse who'd been one of
George Gey's only African-American students.

When I called Roland Pattillo to see what he knew about Henrietta, I told him I was writing a book about her.

"Oh you are?" he said, laughing a slow, rumbling laugh that said,
Oh child, you have no idea what you're getting into. "Henrietta's
family won't talk to you. They've had a terrible time with the HeLa
cells."

"You know her family?" I said. "Can you put me in touch with
them?

"I do have the ability to put you in touch with them, but you need
to answer a few questions, starting with 'Why should I?'"

For the next hour, Pattillo grilled me about my intentions. As I told him about the history of my HeLa obsession, he grumbled and sighed, letting out occasional *mmmmmms* and *welllllll*s.

Eventually he said, "Correct me if I'm wrong, but you *are* white."

"Is it that obvious?"

"Yes," he said. "What do you know about African-Americans and science?"

I told him about the Tuskegee syphilis study like I was giving an oral report in history class: It started in the thirties, when U.S. Public Health Service researchers at the Tuskegee Institute decided to study how syphilis killed, from infection to death. They recruited hundreds of African-American men with syphilis, then watched them die slow, painful, and preventable deaths, even after they realized penicillin could cure them. The research subjects didn't ask questions. They were poor and uneducated, and the researchers offered incentives: free physical exams, hot meals, and rides into town on clinic days, plus fifty-dollar burial stipends for their families when the men died. The researchers chose black subjects because they, like many whites at the time, believed black people were "a notoriously syphilis-soaked race."

The public didn't learn about the Tuskegee study until the seventies, after hundreds of men enrolled in it had already died. The news spread like pox through black communities: doctors were doing research on black people, lying to them, and watching them die. Rumors started circulating that the doctors had actually injected the men with syphilis in order to study them.

"What else?" Pattillo grumbled.

I told him I'd heard about so-called Mississippi Appendectomies, unnecessary hysterectomies performed on poor black women to stop them from reproducing, and to give young doctors a chance to practice the procedure. I'd also read about the lack of funding for research into sickle-cell anemia, a disease that affected blacks almost exclusively.

"It's interesting that you called when you did," he said. "I'm organizing the next HeLa conference, and when the phone rang, I'd just sat down at my desk and typed the words *Henrietta Lacks* on my screen." We both laughed. It must be a sign, we said; perhaps Henrietta wanted us to talk.

"Deborah is Henrietta's baby girl," he said, very matter-of-fact. "The family calls her Dale. She's almost fifty now, still living in Baltimore, with grandchildren of her own. Henrietta's husband is still alive. He's around eighty-four—still goes to the clinics at Johns Hopkins." He dropped this like a tease.

"Did you know Henrietta had an epileptic daughter?" Pattillo asked.

"No."

"She died at fifteen, soon after Henrietta's death. Deborah is the only daughter left," he said. "She came near a stroke recently because of the agony she's gone through regarding inquiries into her mother's death and those cells. I won't be a part of anyone doing that to her again."

I started to speak, but he interrupted me.

"I need to go see patients now," he said abruptly. "I'm not ready to put you in touch with the family yet. But I think you're honest about your intentions. We will talk again after I think. Call back tomorrow."

After three straight days of grilling, Pattillo finally decided to give me Deborah's phone number. But first, he said, there were a few things I needed to know. He lowered his voice and rattled off a list of dos and don'ts for dealing with Deborah Lacks: Don't be aggressive. Do be honest. Don't be clinical, don't try to force her into anything, don't talk down to her, she hates that. Do be compassionate, don't forget that she's been through a lot with these cells, do have patience. "You'll need that more than anything," he told me.

Moments after hanging up the phone with Pattillo, his list of dos and don'ts in my hand, I dialed Deborah's number, then paced as her phone rang. When she whispered hello, I blurted out, "I'm so excited you answered because I've been wanting to talk to you for years! I'm writing a book about your mother!"

"Huh?" she said.

I didn't know that Deborah was nearly deaf—she relied heavily on lip reading and couldn't follow anyone who talked fast.

I took a deep breath and tried again, forcing myself to sound out every syllable.

"Hi, my name is Rebecca."

"How ya doin?" she said, weary but warm.

"I'm very excited to talk to you."

"Mmmhmm," she said, like she'd heard that line many times before.

I told her again that I wanted to write a book about her mother and said I was surprised no one seemed to know anything about her, even though her cells were so important for science.

Deborah sat silent for a long moment, then screamed, "That's right!" She giggled and started talking like we'd known each other for years. "Everything always just about the cells and don't even worry about her name and was HeLa even a person. So hallelujah! I think a book would be great!"

This was not what I'd expected.

I was afraid to say anything that might make her stop talking, so I simply said, "Great." And that was the last word I spoke until the end of our call. I didn't ask a single question, just took notes as fast as I could.

Deborah crammed a lifetime of information into a manic and confusing forty-five minutes that jumped without warning, and in no particular order, from the 1920s to the 1990s, from stories of her father to her grandfather, cousins, mother, and total strangers.

"Nobody never said nothing," she told me. "I mean, where my mother clothes at? Where my mother shoes? I knew about her watch and ring, but it was stolen. That was after my brother killed that boy." She talked about a man she didn't name, saying, "I didn't think it was fit for him to steal my mother medical record and autopsy papers. He was in prison for fifteen years in Alabama. Now he sayin John

Hopkin killed my mother and them white doctors experimented on her cause she was black.

"My nerve broke down," she said. "I just couldn't take it. My speech is coming back a little better—I almost had two strokes in two weeks cause of all that stuff with my mother cells."

Then suddenly she was talking about her family history, saying something about "the Hospital for Crazy Negroes" and her mother's great-grandfather having been a slave owner. "We all mixed. And one of my mother sisters converted to Puerto Rican."

Again and again, she said, "I can't take it anymore," and "Who are we supposed to trust now?" More than anything, she told me, she wanted to learn about her mother and what her cells had done for science. She said people had been promising her information for decades and never delivering it. "I'm sick of it," she said. "You know what I really want? I want to know, what did my mother smell like? For all my life I just don't know anything, not even the little common little things, like what color she like? Did she like to dance? Did she breast-feed me? Lord, I'd like to know that. But nobody ever say nothing."

She laughed and said, "I tell you one thing—the story's not over yet. You got your work cut out for you, girl. This thing's crazy enough for three books!"

Then someone walked through her front door and Deborah yelled straight into the receiver, "Good morning! I got *mail*?" She sounded panicked by the idea of it. "Oh my God! Oh no! Mail?!"

"Okay, Miss Rebecca," she said. "I got to go. You call me Monday, promise? Okay, dear. God bless. Bye-bye."

She hung up and I sat stunned, receiver crooked in my neck, frantically scribbling notes I didn't understand, like *brother = murder, mail = bad, man stole Henrietta's medical records,* and *Hospital for Negro Insane?*

When I called Deborah back as promised, she sounded like a different person. Her voice was monotone, depressed, and slurred, like she was heavily sedated.

"No interviews," she mumbled almost incoherently. "You got to go away. My brothers say I should write my own book. But I ain't a writer. I'm sorry."

I tried to speak, but she cut me off. "I can't talk to you no more. Only thing to do is convince the men." She gave me three phone numbers: her father; her oldest brother, Lawrence; and her brother David Jr.'s pager. "Everybody call him Sonny," she told me, then hung up. I wouldn't hear her voice again for nearly a year.

I started calling Deborah, her brothers, and her father daily, but they didn't answer. Finally, after several days of leaving messages, someone answered at Day's house: a young boy who didn't say hello, just breathed into the receiver, hip-hop thumping in the background.

When I asked for David, the boy said, "Yeah," and threw the phone down.

"Go get Pop!" he yelled, followed by a long pause. "It's important. Get Pop!"

No response.

"Lady's on the phone," he yelled, "come on . . ."

The first boy breathed into the receiver again as a second boy picked up an extension and said hello.

"Hi," I said. "Can I talk to David?"

"Who this?" he asked.

"Rebecca," I said.

He moved the phone away from his mouth and yelled, "Get Pop, lady's on the phone about his wife cells."

Years later I'd understand how a young boy could know why I was calling just from the sound of my voice: the only time white people called Day was when they wanted something having to do with HeLa cells. But at the time I was confused—I figured I must have heard wrong.

A woman picked up a receiver saying, "Hello, may I help you?" She was sharp, curt, like *I do not have time for this.*

I told her I was hoping to talk to David, and she asked who was calling. Rebecca, I said, afraid she'd hang up if I said anything more.

"Just a moment." She sighed and lowered the phone. "Go take this to Day," she told a child. "Tell him he got a long-distance call, somebody named Rebecca calling about his wife cells."

The child grabbed the phone, pressed it to his ear, and ran for Day. Then there was a long silence.

"Pop, get up," the kid whispered. "There's somebody about your wife."

"Whu . . ."

"Get *up*, there's somebody about your wife cells."

"Whu? Where?"

"Wife cells, on the phone . . . get up."

"Where her cells?"

"Here," the boy said, handing Day the phone.

"Yeah?"

"Hi, is this David Lacks?"

"Yeah."

I told him my name and started to explain why I was calling, but before I could say much, he let out a deep sigh.

"Whanowthis," he mumbled in a deep Southern accent, his words slurred like he'd had a stroke. "You got my wife cells?"

"Yeah," I said, thinking he was asking if I was calling about his wife's cells.

"Yeah?" he said, suddenly bright, alert. "You got my wife cells? She know you talking?"

"Yeah," I said, thinking he was asking if Deborah knew I was calling.

"Well, so let my old lady cells talk to you and leave me alone," he snapped. "I had enough 'a you people." Then he hung up.

7

The Death
and Life of
Cell Culture

On April 10, 1951, three weeks after Henrietta started radiation therapy, George Gey appeared on WAAM television in Baltimore for a special show devoted to his work. With dramatic music in the background, the announcer said, "Tonight we will learn why scientists believe that cancer can be conquered."

The camera flashed to Gey, sitting at a desk in front of a wall covered with pictures of cells. His face was long and handsome, with a pointed nose, black plastic bifocals, and a Charlie Chaplin mustache. He sat stiff and straight-backed, tweed suit perfectly pressed, white handkerchief in his breast pocket, hair slicked. His eyes darted off screen, then back to the camera as he drummed his fingers on the desk, his face expressionless.

"The normal cells which make up our bodies are tiny objects, five thousand of which would fit on the head of a pin," he said, his voice a bit too loud and stilted. "How the normal cells become cancerous is still a mystery."

He gave viewers a basic overview of cell structure and cancer using diagrams and a long wooden pointer. He showed films of cells moving across the screen, their edges inching further and further into the empty space around them. And he zoomed in on one cancer cell, its edges round and smooth until it began to quiver and shake violently, exploding into five cancer cells.

At one point he said, "Now let me show you a bottle in which we have grown massive quantities of cancer cells." He picked up a clear glass pint-sized bottle, most likely full of Henrietta's cells, and rocked it in his hands as he explained that his lab was using those cells to find ways to stop cancer. He said, "It is quite possible that from fundamental studies such as these that we will be able to learn a way by which cancer cells can be damaged or completely wiped out."

To help make that happen, Gey began sending Henrietta's cells to any scientist who might use them for cancer research. Shipping live cells in the mail—a common practice today—wasn't done at the time. Instead, Gey sent them via plane in tubes with a few drops of culture medium, just enough to keep them alive for a short time. Sometimes pilots or stewards tucked the tubes in their shirt pockets, to keep the cells at body temperature as if they were still in an incubator. Other times, when the cells had to ride in the cargo hold, Gey tucked them into holes carved in blocks of ice to keep them from overheating, then packed the ice in cardboard boxes filled with sawdust. When shipments were ready to go, Gey would warn recipients that the cells were about to "metastasize" to their cities, so they could stand ready to fetch the shipment and rush back to their labs. If all went well, the cells survived. If not, Gey packaged up another batch and tried again.

He sent shipments of HeLa cells to researchers in Texas, India, New York, Amsterdam, and many places between. Those researchers gave them to more researchers, who gave them to more still. Henrietta's cells rode into the mountains of Chile in the saddlebags of pack mules. As Gey flew from one lab to another, demonstrating his culturing techniques and helping to set up new laboratories, he always flew with tubes of Henrietta's cells in his breast pocket. And when scientists visited Gey's lab to learn his techniques, he usually sent them home with a vial or two of HeLa. In letters, Gey and some of his colleagues began referring to the cells as his "precious babies."

The reason Henrietta's cells were so precious was because they allowed scientists to perform experiments that would have been impossible with a living human. They cut HeLa cells apart and exposed them to endless toxins, radiation, and infections. They bombarded them with drugs, hoping to find one that would kill malignant cells without destroying normal ones. They studied immune suppression and cancer growth by injecting HeLa cells into immune-compromised rats, which developed malignant tumors much like Henrietta's. If the cells died in the process, it didn't matter—scientists could just go back to their eternally growing HeLa stock and start over again.

Despite the spread of HeLa and the flurry of new research that followed, there were no news stories about the birth of the amazing HeLa cell line and how it might help stop cancer. In Gey's one appearance on television, he didn't mention Henrietta or her cells by name, so the general public knew nothing of HeLa. But even if they had known, they probably wouldn't have paid it much mind. For decades the press had been reporting that cell culture was going to save the world from disease and make man immortal, but by 1951 the general public had stopped buying it. Cell culture had become less a medical miracle than something out of a scary science-fiction movie.

It all started on January 17, 1912, when Alexis Carrel, a French surgeon at the Rockefeller Institute, grew his "immortal chicken heart."

Scientists had been trying to grow living cells since before the turn of the century, but their samples had always died. As a result, many researchers believed it was impossible to keep tissues alive outside the body. But Carrel set out to prove them wrong. At age thirty-nine he'd already invented the first technique for suturing blood vessels together, and had used it to perform the first coronary bypass and develop methods for transplanting organs. He hoped someday to grow whole organs in the laboratory, filling massive vaults with lungs, livers, kidneys, and tissues he could ship through the mail for transplantation. As a first step, he'd tried to grow a sliver of chicken-heart tissue in culture, and to everyone's amazement, it worked. Those heart cells kept beating as if they were still in the chicken's body.

Months later, Carrel won a Nobel Prize for his blood-vessel-suturing technique and his contributions to organ transplantation, and he became an instant celebrity. The prize had nothing to do with the chicken heart, but articles about his award conflated the immortal chicken-heart cells with his transplantation work, and suddenly it sounded like he'd found the fountain of youth. Headlines around the world read:

CARREL'S NEW MIRACLE POINTS WAY TO AVERT OLD AGE! . . .
SCIENTISTS GROW IMMORTAL CHICKEN HEART . . .
DEATH PERHAPS NOT INEVITABLE

Scientists said Carrel's chicken-heart cells were one of the most important advances of the century, and that cell culture would uncover the secrets behind everything from eating and sex to "the music of Bach, the poems of Milton, [and] the genius of Michelangelo." Carrel was a scientific messiah. Magazines called his culture medium "an elixir of youth" and claimed that bathing in it might make a person live forever.

But Carrel wasn't interested in immortality for the masses. He was a eugenicist: organ transplantation and life extension were ways to preserve what he saw as the superior white race, which he believed was being polluted by less intelligent and inferior stock, namely the poor, uneducated, and nonwhite. He dreamed of never-ending life for those he deemed worthy, and death or forced sterilization for everyone else. He'd later praise Hitler for the "energetic measures" he took in that direction.

Carrel's eccentricities fed into the media frenzy about his work. He was a stout, fast-talking Frenchman with mismatched eyes—one brown, the other blue—who rarely went out without his surgeon's cap. He wrongly believed that light could kill cell cultures, so his laboratory looked like the photo negative of a Ku Klux Klan rally, where technicians worked in long black robes, heads covered in black hoods with small slits cut for their eyes. They sat on black stools at black tables in a shadowless room with floors, ceilings, and walls painted black. The only illumination came from a small, dust-covered skylight.

Carrel was a mystic who believed in telepathy and clairvoyance, and thought it was possible for humans to live several centuries through the use of suspended animation. Eventually he turned his apartment into a chapel, began giving lectures on medical miracles, and told reporters he dreamed of moving to South America and becoming a dictator. Other researchers distanced themselves, criticizing him for being unscientific, but much of white America embraced his ideas and saw him as a spiritual adviser and a genius.

Reader's Digest ran articles by Carrel advising women that a "husband should not be induced by an oversexed wife to perform a sexual act," since sex drained the mind. In his best-selling book, *Man, the Unknown,* he proposed fixing what he believed was "an error" in the U.S. Constitution that promised equality for all people. "The feeble-minded and the man of genius should not be equal before the law," he wrote. "The stupid, the unintelligent, those who are dispersed, incapable of attention, of effort, have no right to a higher education."

His book sold more than two million copies and was translated into twenty languages. Thousands showed up for Carrel's talks, sometimes requiring police in riot gear to keep order as buildings filled to capacity and fans had to be turned away.

Through all of this, the press and public remained obsessed with Carrel's immortal chicken heart. Each year on New Year's Day, the *New York World Telegram* called Carrel to check on the cells; and every January 17 for decades, when Carrel and his assistants lined up in their black suits to sing "Happy Birthday" to the cells, some newspaper or magazine retold the same story again and again:

CHICKEN HEART CELLS ALIVE TEN YEARS . . .
FOURTEEN YEARS . . . TWENTY . . .

Each time, the stories promised the cells would change the face of medicine, but they never did. Meanwhile, Carrel's claims about the cells grew more fantastical.

At one point he said the cells "would reach a volume greater than that of the solar system." *The Literary Digest* reported that the cells could have already "covered the earth," and a British tabloid said they could "form a rooster . . . big enough today to cross the Atlantic in a single stride, [a bird] so monstrous that when perched on this mundane sphere, the world, it would look like a weathercock." A string of best-selling books warned of the dangers of tissue culture: one predicted that 70 percent of babies would soon be grown in culture; another imagined tissue culture producing giant "Negroes" and two-headed toads.

But the fear of tissue culture truly found its way into American living rooms in an episode of *Lights Out,* a 1930s radio horror show that told the story of a fictional Dr. Alberts who'd created an immortal chicken heart in his lab. It grew out of control, filling the city streets like The Blob, consuming everyone and everything in its path. In only two weeks it destroyed the entire country.

The real chicken-heart cells didn't fare so well. In fact, it turned out that the original cells had probably never survived long at all. Years after Carrel died awaiting trial for collaborating with the Nazis, scientist Leonard Hayflick grew suspicious of the chicken heart. No one had ever been able to replicate Carrel's work, and the cells seemed to defy a basic rule of biology: that normal cells can only divide a finite number of times before dying. Hayflick investigated them and concluded that the original chicken-heart cells had actually died soon after Carrel put them in culture, and that, intentionally or not, Carrel had been putting new cells in the culture dishes each time he "fed" them using an "embryo juice" he made from ground tissues. At least one of Carrel's former lab assistants verified Hayflick's suspicion. But no one could test the theory, because two years after Carrel's death, his assistant unceremoniously threw the famous chicken-heart cells in the trash.

Either way, by 1951, when Henrietta Lacks's cells began growing in the Gey lab—just five years after the widely publicized "death" of Carrel's chicken heart—the public image of immortal cells was tarnished. Tissue culture was the stuff of racism, creepy science fiction, Nazis, and snake oil. It wasn't something to be celebrated. In fact, no one paid much attention to it at all.

8

"A Miserable Specimen"

In early June, Henrietta told her doctors several times that she thought the cancer was spreading, that she could feel it moving through her, but they found nothing wrong with her. "The patient states that she feels fairly well," one doctor wrote in her chart, "however she continues to complain of some vague lower abdominal discomfort. . . . No evidence of recurrence. Return in one month."

There's no indication that Henrietta questioned him; like most patients in the 1950s, she deferred to anything her doctors said. This was a time when "benevolent deception" was a common practice— doctors often withheld even the most fundamental information from their patients, sometimes not giving them any diagnosis at all. They believed it was best not to confuse or upset patients with frightening terms they might not understand, like *cancer*. Doctors knew best, and most patients didn't question that.

Especially black patients in public wards. This was 1951 in Baltimore, segregation was law, and it was understood that black people didn't question white people's professional judgment. Many black patients were just glad to be getting treatment, since discrimination in hospitals was widespread.

There's no way of knowing whether or how Henrietta's treatment would have differed if she'd been white. According to Howard Jones, Henrietta got the same care any white patient would have; the biopsy, the radium treatment, and radiation were all standard for the day. But several studies have shown that black patients were treated and hospitalized at later stages of their illnesses than white patients. And once hospitalized, they got fewer pain medications, and had higher mortality rates.

All we can know for sure are the facts of Henrietta's medical records: a few weeks after the doctor told her she was fine, she went back to Hopkins saying that the "discomfort" she'd complained about last time was now an "ache" in both sides. But the doctor's entry was identical to the one weeks earlier: "No evidence of recurrence. Return in one month."

Two and a half weeks later, Henrietta's abdomen hurt, and she could barely urinate. The pain made it hard to walk. She went back to Hopkins, where a doctor passed a catheter to empty her bladder, then sent her home. Three days later, when she returned complaining once again of pain, a doctor pressed on her abdomen and felt a "stony hard" mass. An X-ray showed that it was attached to her pelvic wall, nearly blocking her urethra. The doctor on duty called for Jones and several others who'd treated Henrietta; they all examined her and looked at the X-ray. "Inoperable," they said. Only weeks after a previous entry declared her healthy, one of the doctors wrote, "The patient looks chronically ill. She is obviously in pain." He sent her home to bed.

Sadie would later describe Henrietta's decline like this: "Hennie didn't fade away, you know, her looks, her body, it didn't just fade. Like some peoples be sick in the bed with cancer and they look so *bad*. But she didn't. The only thing you could tell was in her eyes. Her eyes were tellin you that she wasn't gonna be alive no more."

Until that point, no one except Sadie, Margaret, and Day knew Henrietta was sick. Then, suddenly, everyone knew. When Day and the cousins walked home from Sparrows Point after each shift, they could hear Henrietta from a block away, wailing for the Lord to help her. When Day drove her back to Hopkins for X-rays the following week, stone-hard tumors filled the inside of her abdomen: one on her uterus, one on each kidney and on her urethra. Just a month after a note in her medical record said she was fine, another doctor wrote, "In view of the rapid extension of the disease process the outlook is quite poor." The only option, he said, was "further irradiation in the hopes that we may at least relieve her pain."

Henrietta couldn't walk from the house to the car, but either Day or one of the cousins managed to get her to Hopkins every day for radiation. They didn't realize she was dying. They thought the doctors were still trying to cure her.

Each day, Henrietta's doctors increased her dose of radiation, hoping it would shrink the tumors and ease the pain until her death. Each day the skin on her abdomen burned blacker and blacker, and the pain grew worse.

On August 8, just one week after her thirty-first birthday, Henrietta arrived at Hopkins for her treatment, but this time she said she wanted to stay. Her doctor wrote, "Patient has been complaining bitterly of pain and she seems genuinely miserable. She has to come in from a considerable distance and it is felt that she deserves to be in the hospital where she can be better cared for."

After Henrietta checked into the hospital, a nurse drew blood and labeled the vial COLORED, then stored it in case Henrietta needed transfusions later. A doctor put Henrietta's feet in stirrups once again, to take a few more cells from her cervix at the request of George Gey, who wanted to see if a second batch would grow like the first. But Henrietta's body had become so contaminated with toxins normally flushed from the system in urine, her cells died immediately in culture.

During Henrietta's first few days in the hospital, the children came with Day to visit her, but when they left, she cried and moaned for hours. Soon the nurses told Day he couldn't bring the children anymore, because it upset Henrietta too much. After that, Day would park the Buick behind Hopkins at the same time each day and sit on a little patch of grass on Wolfe Street with the children, right under Henrietta's window. She'd pull herself out of bed, press her hands and face to the glass, and watch her children play on the lawn. But within days, Henrietta couldn't get herself to the window anymore.

Her doctors tried in vain to ease her suffering. "Demerol does not seem to touch the pain," one wrote, so he tried morphine. "This doesn't help too much either." He gave her Dromoran. "This stuff works," he wrote. But not for long. Eventually one of her doctors tried injecting pure alcohol straight into her spine. "Alcohol injections ended in failure," he wrote.

New tumors seemed to appear daily—on her lymph nodes, hip bones, labia—and she spent most days with a fever up to 105. Her doctors stopped the radiation treatment and seemed as defeated by the cancer as she was. "Henrietta is still a miserable specimen," they wrote. "She groans." "She is constantly nauseated and claims she vomits everything she eats." "Patient acutely upset . . . very anxious." "As far as I can see we are doing all that can be done."

There is no record that George Gey ever visited Henrietta in the hospital, or said anything to her about her cells. And everyone I talked to who might know said that Gey and Henrietta never met. Everyone, that is, except Laure Aurelian, a microbiologist who was Gey's colleague at Hopkins.

"I'll never forget it," Aurelian said. "George told me he leaned over Henrietta's bed and said, 'Your cells will make you immortal.' He told Henrietta her cells would help save the lives of countless people, and she smiled. She told him she was glad her pain would come to some good for someone."

9

Turner Station

A few days after my first conversation with Day, I drove from Pittsburgh to Baltimore to meet his son, David "Sonny" Lacks Jr. He'd finally called me back and agreed to meet, saying he'd gotten worn out from my number showing up on his pager. I didn't know it then, but he'd made five panicked phone calls to Pattillo, asking questions about me before calling.

The plan was that I'd page Sonny when I got to Baltimore, then he'd pick me up and take me to his brother Lawrence's house to meet their father and—if I was lucky—Deborah. So I checked in to the downtown Holiday Inn, sat on the bed, phone in my lap, and dialed Sonny's pager. No reply.

I stared through my hotel room window at a tall, Gothic-looking brick tower across the street with a huge clock at the top. It was a weatherbeaten silver, with big letters spelling B-R-O-M-O-S-E-L-T-Z-E-R in a circle around its face. I watched the hands move slowly past the letters, paged Sonny every few minutes, and waited for the phone to ring.

Eventually I grabbed the fat Baltimore phone book, opened to the

*L*s, and ran my finger down a long line of names: *Annette Lacks* . . . *Charles Lacks* . . . I figured I'd call every Lacks in the book asking if they knew Henrietta. But I didn't have a cell phone and didn't want to tie up the line, so I paged Sonny again, then lay back on the bed, phone and White Pages still in my lap. I started rereading a yellowed copy of a 1976 *Rolling Stone* article about the Lackses by a writer named Michael Rogers—the first reporter ever to contact Henrietta's family. I'd read it many times, but wanted every word fresh in my mind.

Halfway through the article, Rogers wrote, "I am sitting on the seventh floor of the downtown Baltimore Holiday Inn. Through the thermopane picture window is a huge public clock in which the numerals have been replaced by the characters B-R-O-M-O-S-E-L-T-Z-E-R; in my lap is a telephone, and the Baltimore White Pages."

I bolted upright, suddenly feeling like I'd been sucked into a *Twilight Zone* episode. More than two decades earlier—when I was just three years old—Rogers had gone through those same White Pages. "Halfway through the 'Lacks' listings it becomes clear that just about everybody had known Henrietta," he wrote. So I opened the phone book again and started dialing, hoping I'd find one of those people who knew her. But they didn't answer their phones, they hung up on me, or they said they'd never heard of Henrietta. I dug out an old newspaper article where I'd seen Henrietta's Turner Station address: 713 New Pittsburgh Avenue. I looked at four maps before finding one where Turner Station wasn't covered by ads or blow-up grids of other neighborhoods.

It turned out Turner Station wasn't just hidden on the map. To get there, I had to drive past the cement wall and fence that blocked it from the interstate, across a set of tracks, past churches in old storefronts, rows of boarded-up houses, and a buzzing electrical generator as big as a football field. Finally I saw a dark wooden sign saying WELCOME TO TURNERS STATION in the parking lot of a fire-scorched bar with pink tasseled curtains.

To this day no one's entirely sure what the town is actually called, or how to spell it. Sometimes it's plural (Turners Station), other times possessive (Turner's Station), but most often it's singular (Turner

Station). It was originally deeded as "Good Luck," but never quite lived up to the name.

When Henrietta arrived there in the forties, the town was booming. But the end of World War II brought cutbacks at Sparrows Point. Baltimore Gas and Electric demolished three hundred homes to make room for a new power plant, leaving more than 1,300 homeless, most of them black. More and more land was zoned for industrial use, which meant more houses torn down. People fled for East Baltimore or back to the country, and the population of Turner Station dropped by half before the end of the fifties. By the time I got there, it was about one thousand and falling steadily, because there were few jobs.

In Henrietta's day, Turner Station was a town where you never locked your doors. Now there was a housing project surrounded by a 13,000-foot-long brick-and-cement security wall in the field where Henrietta's children once played. Stores, nightclubs, cafés, and schools had closed, and drug dealers, gangs, and violence were on the rise. But Turner Station still had more than ten churches.

The newspaper article where I'd gotten Henrietta's address quoted a local woman, Courtney Speed, who owned a grocery store and had created a foundation devoted to building a Henrietta Lacks museum. But when I got to the lot where Speed's Grocery was supposed to be, I found a gray, rust-stained mobile home, its broken windows covered with wire. The sign out front had a single red rose painted on it, and the words REVIVING THE SPIRIT TO RECAPTURE THE VISION. PROVERBS 29:18. Six men gathered on the front steps, laughing. The oldest, in his thirties, wore red slacks, red suspenders, a black shirt, and a driving cap. Another wore an oversized red and white ski jacket. They were surrounded by younger men of various shades of brown in sagging pants. The two men in red stopped talking, watched me drive by slowly, then kept on laughing.

Turner Station is less than a mile across in any direction, its horizon lined with skyscraper-sized shipping cranes and smokestacks billowing thick clouds from Sparrows Point. As I drove in circles looking for Speed's Grocery, children stopped playing in the streets to stare and wave. They ran between matching red-brick houses and past women hanging fresh laundry, following me as their mothers smiled and waved too.

I drove by the trailer with the men out front so many times, they started waving at me with each pass. I did the same with Henrietta's old house. It was a unit in a brown brick building divided into four homes, with a chain-link fence, several feet of grass out front, and three steps leading up to a small cement stoop. A child watched me from behind Henrietta's old screen door, waving and playing with a stick.

I waved back at everyone and feigned surprise each time the group of children following me appeared on various streets grinning, but I didn't stop and ask for help. I was too nervous. The people of Turner Station just watched me, smiling and shaking their heads like, *What's that young white girl doing driving around in circles?*

Finally I saw the New Shiloh Baptist Church, which the newspaper article had mentioned as the site of community meetings about the Henrietta Lacks museum. But it was closed. As I pressed my face to the tall glass out front, a black town car pulled up, and a smooth, handsome man in his forties jumped out, with gold-tinted glasses, black suit, black beret, and the keys to the church. He slid his glasses to the end of his nose and looked me over, asking if I needed help.

I told him why I was there.

"Never heard of Henrietta Lacks," he said.

"Not many people have," I said, and told him I'd read that someone had hung a plaque in Henrietta's honor at Speed's Grocery.

"Oh! Speed's?" he said, suddenly all smiles and a hand on my shoulder. "I can take you to Speed's!" He told me to get in my car and follow him.

Everyone on the street waved and yelled as we passed: "Hi Reverend Jackson!" "How you doin, Reverend?" He nodded and yelled right back, "How you doin!" "God bless you!" Just two blocks away, we stopped in front of that gray trailer with the men out front and the Reverend jammed his car into park, waving for me to get out. The cluster of men on the steps smiled, grabbed the pastor's hand, and gave it two-handed shakes, saying, "Hey Reverend, you brought a friend?"

"Yes I did," he told them. "She's here to talk to Ms. Speed."

The one in the red pants and red suspenders—who turned out to be Speed's oldest son, Keith—said she was out, and who knew when she'd be back, so I may as well grab a seat on the porch with the boys and wait. As I sat down, the man in the red and white ski jacket smiled a big bright smile, then told me he was her son Mike. Then there were her sons Cyrus and Joe and Tyrone. Every man on that porch was her son; so was nearly every man that walked in the store. Pretty soon, I'd counted fifteen sons and said, "Wait a minute. She's got fifteen kids?"

"Oh!" Mike yelled. "You don't know Mama Speed, do you?! Oooh, I look up to Mama—she tough! She keep Turners Station in line, boy! She fears no man!"

The men on the porch all nodded and said, "That's right."

"Don't you get scared if anybody come in here try to attack Mama when we're not around," Mike said, "cause she'll scare them to death!" Speed's sons let out a chorus of amens as Mike told a story, saying, "This man came in the store once yellin, 'I'm gonna come cross that counter and get you.' I was hidin behind Mama I was so scared! And do you know what Mama did? She rocked her head and raised up them arms and said, 'Come on! Come onnnnn! If you think you crazy, you just try it!'"

Mike slapped me on the back and all the sons laughed.

At that moment, Courtney Speed appeared at the bottom of the steps, her long black hair piled loose on her head, strands hanging in wisps around her face, which was thin, beautiful, and entirely ageless. Her eyes were soft brown with a perfect halo of sea blue around the edges. She was delicate, not a hard edge on her. She hugged a grocery bag to her chest and whispered, "But did that man jump across that counter at me?"

Mike screamed and laughed so hard he couldn't answer.

She looked at him, calm and smiling. "I said, *Did that man jump?*"

"No, he did not!" Mike said, grinning. "That man didn't do

nuthin but run! That's why Mama got no gun in this store. She don't need one!"

"I don't live by the gun," she said, then turned to me and smiled. "How *you* doin?" She walked up the stairs into the store, and we all followed.

"Mama," Keith said, "Pastor brought this woman in here. She's Miss Rebecca and she's here to talk to you."

Courtney Speed smiled a beautiful, almost bashful smile, her eyes bright and motherly. "God bless you, sweetie," she said.

Inside, flattened cardboard boxes covered most of the floor, which was worn from years of foot traffic. Shelves lined each wall, some bare, others stacked with Wonder Bread, rice, toilet paper, and pigs' feet. On one, Speed had piled hundreds of editions of the *Baltimore Sun* dating back to the 1970s, when her husband died. She said she'd given up replacing the windows each time someone broke in because they'd just do it again. She'd hung handwritten signs on every wall of the store: one for "Sam the Man Snowballs," others for sports clubs, church groups, and free GED and adult literacy classes. She had dozens of "spiritual sons," who she treated no different than her six biological sons. And when any child came in to buy chips, candy, or soda, Speed made them calculate how much change she owed them—they got a free Hershey's kiss for each correct answer.

Speed started straightening the items on her shelves so each label faced out, then yelled over her shoulder at me, "How did you find your way here?"

I told her about the four maps, and she threw a box of lard onto the shelf. "Now we got the four-map syndrome," she said. "They keep trying to push us off the earth, but God won't let them. Praise the Lord, he brings us the people we really need to talk to."

She wiped her hands on her white shirt. "Now that He brought you here, what can I do for you?"

"I'm hoping to learn about Henrietta Lacks," I said.

Courtney gasped, her face suddenly ashen. She took several steps back and hissed, "You know Mr. Cofield? Did he send you?"

I was confused. I told her I'd never heard of Cofield, and no one had sent me.

"How did you know about me?" she snapped, backing away further.

I pulled the old crumpled newspaper article from my purse and handed it to her.

"Have you talked to the family?" she asked.

"I'm trying," I said. "I talked to Deborah once, and I was supposed to meet Sonny today, but he didn't show up."

She nodded, like *I knew it.* "I can't tell you anything until you got the support of the family. I can't risk that."

"What about the plaque you got for the museum?" I asked. "Can I see that?"

"It's not here," she snapped. "Nothing's here, because bad things happened around all that."

She looked at me for a long moment, then her face softened. She took my hand in one of hers, and touched my face with the other.

"I like your eyes," she said. "Come with me."

She hurried out the door and down the stairs to her old brown station wagon. A man sat in the passenger seat, staring straight at the road as if the car were moving. He didn't look up as she jumped in, saying, "Follow me."

We drove through Turner Station to the parking lot of the local public library. As I opened my car door, Courtney appeared, clapping, grinning, and bouncing on her tiptoes. Words erupted from her: "February first is Henrietta Lacks day here in Baltimore County," she said. "This February first is going to be the big kickoff event here at the library! We're still trying to put a museum together, even though the Cofield situation did cause so many problems. Terrified Deborah. We were supposed to be almost done with the museum by now—we were so close before all that horribleness. But I'm glad He sent you," she said, pointing to the sky. "This story just *got* to be told! Praise the Lord, people *got* to know about Henrietta!"

"Who's Cofield?" I asked.

She cringed and slapped her hand over her mouth. "I really can't talk until the family says it's okay," she said, then grabbed my hand and ran into the library.

"This is Rebecca," she told the librarian, bouncing on her toes again. "She's writing about Henrietta Lacks!"

"Oh, that's wonderful!" the librarian said. Then she looked at Courtney. "Are you talking to her?"

"I need the tape," Courtney said.

The librarian walked down a row of videos, pulled a white box from the shelf, and handed it to her.

Courtney tucked the video under her arm, grabbed my hand, and ran me back to the parking lot, where she jumped into her car and sped off, waving for me to follow. We stopped outside a convenience store while the man in her front seat got out and bought a loaf of bread. Then we dropped him off in front of his house as Courtney yelled back to me, "He's my deaf cousin! Can't drive!"

Finally she led me to a small beauty parlor she owned, not far from Speed's Grocery. She unlocked two bolts on the front door and waved her hand in the air, saying, "Smells like I got a mouse in one of those traps." The shop was narrow, with barber chairs lining one wall and dryers along the other. The hair-washing sink, propped up with a piece of plywood, drained into a large white bucket, the walls around it splattered with years' worth of hair dye. Next to the sink sat a price board: Cut and style ten dollars. Press and curl, seven. And against the back wall, on top of a supply cabinet, sat a photocopy of the picture of Henrietta Lacks, hands on hips, in a pale wood frame several inches too big.

I pointed to the photo and raised my eyebrows. Courtney shook her head.

"I'll tell you everything I know," she whispered, "just as soon as you talk to the family and they say it's okay. I don't want any more problems. And I don't want Deborah to get sick over it again."

She pointed to a cracked red vinyl barber's chair, which she spun to face a small television next to the hair dryers. "You have to watch this tape," she said, handing me the remote and a set of keys. She started to walk out the door, then turned. "Don't you open this door for nothing or nobody but me, you hear?" she said. "And don't you miss nothing in that video—use that rewind button, watch it twice if you have to, but don't you miss nothing."

Then she left, locking the door behind her.

What rolled in front of me on that television screen was a one-hour BBC documentary about Henrietta and the HeLa cells, called *The Way of All Flesh,* which I'd been trying to get a copy of for months. It opened to sweet music and a young black woman who wasn't Henrietta, dancing in front of the camera. A British man began narrating, his voice melodramatic, like he was telling a ghost story that just might be true.

"In 1951 a woman died in Baltimore in America," he said, pausing for effect. "She was called Henrietta Lacks." The music grew louder and more sinister as he told the story of her cells: "These cells have transformed modern medicine. . . . They shaped the policies of countries and of presidents. They even became involved in the Cold War. Because scientists were convinced that in her cells lay the secret of how to conquer death. . . ."

What really grabbed me was footage of Clover, an old plantation town in southern Virginia, where some of Henrietta's relatives still seemed to live. The last image to appear on the screen was Henrietta's cousin Fred Garret, standing behind an old slave shack in Clover, his back to the family cemetery where the narrator said Henrietta lay buried in an unmarked grave.

Fred pointed to the cemetery and looked hard into the camera.

"Do you think them cells still livin?" he asked. "I talkin bout in the grave." He paused, then laughed a long, rumbling laugh. "Hell naw," he said, "I don't guess they are. But they're still livin out in the test tubes. That's a miracle."

The screen went blank and I realized, if Henrietta's children and husband wouldn't talk to me, I needed to visit Clover and find her cousins.

That night, back at the hotel, I finally got Sonny on the phone. He said he'd decided not to meet me but wouldn't tell me why. When I asked him to put me in touch with his family in Clover, he told me to go there and find them myself. Then he laughed and wished me luck.

10

The Other
Side of the
Tracks

Clover sits a few rolling hills off Route 360 in southern Virginia, just past Difficult Creek on the banks of the River of Death. I pulled into town under a blue December sky, with air warm enough for May, a yellow Post-it note with the only information Sonny had given me stuck on my dashboard: "They haven't found her grave. Make sure it's day—there are no lights, gets darker than dark. Ask anybody where Lacks Town is."

Downtown Clover started at a boarded-up gas station with RIP spray-painted across its front, and ended at an empty lot that once held the depot where Henrietta caught her train to Baltimore. The roof of the old movie theater on Main Street had caved in years ago, its screen landing flat in a field of weeds. The other businesses looked like someone left for lunch decades earlier and never bothered coming back: one wall of Abbott's clothing store was lined with boxes of new Red Wing work boots stacked to the ceiling and covered in thick dust; inside its long glass counter, beneath an antique cash register, lay rows and rows of men's dress shirts, still folded starch-stiff in their plastic. The lounge at Rosie's restaurant was filled with overstuffed chairs, couches, and shag carpet, all in dust-covered browns, oranges, and yellows. A sign in the front window said OPEN 7 DAYS, just above one that said CLOSED. At Gregory and Martin Super Market, half-full shopping carts rested in the aisles next to decades-old canned foods, and the wall clock hadn't moved past 6:34 since Martin closed up shop to become an undertaker sometime in the eighties.

Even with kids on drugs and the older generation dying off, Clover didn't have enough death to keep an undertaker in business: in 1974 it had a population of 227; in 1998 it was 198. That same year, Clover lost its town charter. It did still have several churches and a few beauty parlors, but they were rarely open. The only steady business left downtown was the one-room brick post office, but it was closed when I got there.

Main Street felt like a place where you could sit for hours without seeing a pedestrian or a car. But a man stood in front of Rosie's, leaning against his red motorized bicycle, waiting to wave at any cars that might pass. He was a short, round white man with red cheeks who could have been anywhere from fifty to seventy. Locals called him the Greeter, and he'd spent most of his life on that corner waving at anyone who drove by, his face expressionless. I asked if he could direct me to Lacks Town, where I planned to look for mailboxes with the name Lacks on them, then knock on doors asking about Henrietta. The man never said a word, just waved at me, then slowly pointed behind him, across the tracks.

The dividing line between Lacks Town and the rest of Clover was stark. On one side of the two-lane road from downtown, there were vast, well-manicured rolling hills, acres and acres of wide-open property with horses, a small pond, a well-kept house set back from the road, a minivan, and a white picket fence. Directly across the street stood a small one-room shack about seven feet wide and twelve feet long; it was made of unpainted wood, with large gaps between the wallboards where vines and weeds grew.

That shack was the beginning of Lacks Town, a single road about a mile long and lined with dozens of houses—some painted bright yellows or greens, others unpainted, half caved-in or nearly burnt-down. Slave-era cabins sat next to cinder-block homes and trailers, some with satellite dishes and porch swings, others rusted and half buried. I drove the length of Lacks Town Road again and again, past the END OF STATE MAINTENANCE sign where the road turned to gravel, past a tobacco field with a basketball court in it—just a patch of red dirt and a bare hoop attached to the top of a weathered tree trunk.

The muffler on my beat-up black Honda had fallen off some-where between Pittsburgh and Clover, which meant everyone in Lacks Town heard each time I passed. They walked onto porches and peered through windows as I drove by. Finally, on my third or fourth pass, a man who looked like he was in his seventies shuffled out of a green two-room wooden cabin wearing a bright green sweater, a matching scarf, and a black driving cap. He waved a stiff arm at me, eyebrows raised.

"You lost?" he yelled over my muffler.

I rolled down my window and said not exactly.

"Well where you tryin to go?" he said. "Cause I know you're not from around here."

I asked him if he'd heard of Henrietta.

He smiled and introduced himself as Cootie, Henrietta's first cousin.

His real name was Hector Henry—people started calling him Cootie when he got polio decades earlier; he was never sure why. Cootie's skin was light enough to pass for Latino, so when he got sick at nine years old, a local white doctor snuck him into the nearest hos-pital, saying Cootie was his son, since the hospitals didn't treat black patients. Cootie spent a year inside an iron lung that breathed for him, and he'd been in and out of hospitals ever since.

The polio had left him partially paralyzed in his neck and arms, with nerve damage that caused constant pain. He wore a scarf regard-less of the weather, because the warmth helped ease the pain.

I told him why I was there, and he pointed up and down the road. "Everybody in Lacks Town kin to Henrietta, but she been gone so long, even her memory pretty much dead now," he said. "Everything about Henrietta dead except them cells."

He pointed to my car. "Turn this loud thing off and come inside. I'll fix you some juice."

His front door opened into a tiny kitchen with a coffeemaker, a vintage toaster, and an old woodstove with two cooking pots on top, one empty, the other filled with chili. He'd painted the kitchen walls the same dark olive green as the outside, and lined them with power strips and fly swatters. He'd recently gotten indoor plumbing, but still preferred the outhouse.

Though Cootie could barely move his arms, he'd built the house on his own, teaching himself construction as he went along, hammering the plywood walls and plastering the inside. But he'd forgotten to use insulation, so soon after he finished it, he tore down the walls and started over again. A few years after that, the whole place burned down when he fell asleep under an electric blanket, but he built it back up again. The walls were a bit crooked, he said, but he'd used so many nails, he didn't think it would ever fall down.

Cootie handed me a glass of red juice and shooed me out of the kitchen into his dark, wood-paneled living room. There was no couch, just a few metal folding chairs and a barber's chair anchored to the linoleum floor, its cushions covered entirely with duct tape. Cootie had been the Lacks Town barber for decades. "That chair cost twelve hundred dollars now, but I got it for eight dollars back then," he yelled from the kitchen. "Haircut wasn't but a dollar—sometimes I cut fifty-eight heads in one day." Eventually he quit because he couldn't hold his arms up long enough to cut.

A small boom box leaned against one wall blaring a gospel call-in show, with a preacher screaming something about the Lord curing a caller of hepatitis.

Cootie opened a folding chair for me, then walked into his bedroom. He lifted his mattress with one arm, propped it on his head, and began rummaging through piles of paper hidden beneath it.

"I know I got some information on Henrietta in here somewhere," he mumbled from under the mattress. "Where the hell I put that . . . You know other countries be buying her for twenty-five dollars, sometimes fifty? Her family didn't get no money out of it."

After digging through what looked like hundreds of papers, he came back to the living room.

"This here the only picture I got of her," he said, pointing to a copy of the *Rolling Stone* article with the ever-present hands-on-hips photo. "I don't know what it say. Only education I got, I had to learn on my own. But I always couldn't count, and I can't hardly read or write my name cause my hand's so jittery." He asked if the article said anything about her childhood in Clover. I shook my head no.

"Everybody liked Henrietta cause she was a very good condition person," he said. "She just lovey dovey, always smilin, always takin care of us when we come to the house. Even after she got sick, she never was a person who say 'I feel bad and I'm going to take it out on you.' She wasn't like that, even when she hurtin. But she didn't seem to understand what was going on. She didn't want to think she was gonna die."

He shook his head. "You know, they said if we could get all the pieces of her together, she'd weigh over eight hundred pounds now," he told me. "And Henrietta never was a big girl. She just still growin."

In the background, the radio preacher screamed "Hallelujah!" over and over as Cootie spoke.

"She used to take care of me when my polio got bad," he told me. "She always did say she wanted to fix it. She couldn't help me cause I had it before she got sick, but she saw how bad it got. I imagine that's why she used them cells to help get rid of it for other folk." He paused. "Nobody round here never understood how she dead and that thing still livin. That's where the mystery's at."

He looked around the room, nodding his head toward spaces between the wall and ceiling where he'd stuffed dried garlic and onions.

"You know, a lot of things, they man-made," he told me, dropping his voice to a whisper. "You know what I mean by *man-made*, don't you?"

I shook my head no.

"Voodoo," he whispered. "Some peoples is sayin Henrietta's sickness and them cells was man- or woman-made, others say it was doctor-made."

As he talked, the preacher's voice on the radio grew louder, saying, "The Lord, He's gonna help you, but you got to call me right now. If my daughter or sister had cancer! I would get on that phone, cause time's running out!"

Cootie yelled over the radio. "Doctors say they never heard of another case like Henrietta's! I'm sure it was either man-made or spirit-made, one of the two."

Then he told me about spirits in Lacks Town that sometimes visited people's houses and caused disease. He said he'd seen a man spirit in his house, sometimes leaning against the wall by his woodstove, other times by the bed. But the most dangerous spirit, he told me, was the several-ton headless hog he saw roaming Lacks Town years ago with no tail. Links of broken chain dangled from its bloodstained neck, dragging along dirt roads and clanking as it walked.

"I saw that thing crossin the road to the family cemetery," Cootie told me. "That spirit stood right there in the road, its chain swingin and swayin in the breeze." Cootie said it looked at him and stomped its foot, kicking red dust all around its body, getting ready to charge. Just then, a car came barreling down the road with only one headlight.

"The car came along, shined a light right on it, I swear it was a hog," Cootie said. Then the spirit vanished. "I can still hear that chain draggin." Cootie figured that car saved him from getting some new disease.

"Now I don't know for sure if a spirit got Henrietta or if a doctor did it," Cootie said, "but I do know that her cancer wasn't no regular cancer, cause regular cancer don't keep on growing after a person die."

11

"The Devil of
Pain Itself"

By September, Henrietta's body was almost entirely taken over by tumors. They'd grown on her diaphragm, her bladder, and her lungs. They'd blocked her intestines and made her belly swell like she was six months pregnant. She got one blood transfusion after another because her kidneys could no longer filter the toxins from her blood, leaving her nauseated from the poison of her own body. She got so much blood that one doctor wrote a note in her record stopping all transfusions "until her deficit with the blood bank was made up."

When Henrietta's cousin Emmett Lacks heard somebody at Sparrows Point say Henrietta was sick and needed blood, he threw down the steel pipe he was cutting and ran looking for his brother and some friends. They were working men, with steel and asbestos in their lungs and years' worth of hard labor under their calluses and cracked fingernails. They'd all slept on Henrietta's floor and eaten her spaghetti when they first came to Baltimore from the country, and anytime money ran low. She'd ridden the streetcar to and from Sparrows Point to make sure they didn't get lost during their first weeks in the city.

She'd packed their lunches until they found their feet, then sent extra food to work with Day so they didn't go hungry between paychecks. She'd teased them about needing wives and girlfriends, and sometimes helped them find good ones. Emmett had stayed at Henrietta's so long, he had his own bed in the hallway at the top of the stairs. He'd only moved out a few months earlier.

The last time Emmett saw Henrietta, he'd taken her to visit Elsie in Crownsville. They found her sitting behind barbed wire in the corner of a yard outside the brick barracks where she slept. When she saw them coming she made her birdlike noise, then ran to them and just stood, staring. Henrietta wrapped her arms around Elsie, looked her long and hard in the eyes, then turned to Emmett.

"She look like she doin better," Henrietta said. "Yeah, Elsie look nice and clean and everything." They sat in silence for a long time. Henrietta seemed relieved, almost desperate, to see Elsie looking okay. That was the last time she would see her daughter—Emmett figures she knew she was saying goodbye. What she didn't know was that no one would ever visit Elsie again.

A few months later, when Emmett heard Henrietta needed blood, he and his brother and six friends piled into a truck and went straight to Hopkins. A nurse led them through the colored ward, past rows of hospital beds to the one where Henrietta lay. She'd withered from 140 pounds to about 100. Sadie and Henrietta's sister Gladys sat beside her, their eyes swollen from too much crying and not enough sleep. Gladys had come from Clover by Greyhound as soon as she got word Henrietta was in the hospital. The two had never been close, and people still teased Gladys, saying she was too mean and ugly to be Henrietta's sister. But Henrietta was family, so Gladys sat beside her, clutching a pillow in her lap.

A nurse stood in the corner watching as the eight big men crowded around the bed. When Henrietta tried to move her arm to lift herself, Emmett saw the straps around her wrists and ankles, attaching her to the bed frame.

"What you doin here?" Henrietta moaned.

"We come to get you well," Emmett said to a chorus of yeahs from the other men.

Henrietta didn't say a word. She just lay her head back on the pillow.

Suddenly her body went rigid as a board. She screamed as the nurse ran to the bed, tightening the straps around Henrietta's arms and legs to keep her from thrashing onto the floor as she'd done many times before. Gladys thrust the pillow from her lap into Henrietta's mouth, to keep her from biting her tongue as she convulsed in pain. Sadie cried and stroked Henrietta's hair.

"Lord," Emmett told me years later. "Henrietta rose up out that bed wailin like she been possessed by the devil of pain itself."

The nurse shooed Emmett and his brothers out of the ward to the room designated for colored blood collection, where they'd donate eight pints of blood. As Emmett walked from Henrietta's bedside, he turned to look just as the fit began to pass and Gladys slid the pillow from Henrietta's mouth.

"That there's a memory I'll take to my grave," he told me years later. "When them pains hit, looked like her mind just said, *Henrietta, you best leave.* She was sick like I never seen. Sweetest girl you ever wanna meet, and prettier than anything. But them cells, boy, them cells of hers is somethin else. No wonder they never could kill them . . . That cancer was a terrible thing."

Soon after Emmett and his friends visited, at four o'clock on the afternoon of September 24, 1951, a doctor injected Henrietta with a heavy dose of morphine and wrote in her chart, "Discontinue all medications and treatments except analgesics." Two days later, Henrietta awoke terrified, disoriented, wanting to know where she was and what the doctors had been doing to her. For a moment she forgot her own name. Soon after that, she turned to Gladys and told her she was going to die.

"You make sure Day takes care of them children," Henrietta told her sister, tears streaming down her face. "Especially my baby girl Deborah." Deborah was just over a year old when Henrietta went into the hospital. Henrietta had wanted to hold Deborah, to dress her in beautiful clothes and braid her hair, to teach her how to paint her nails, curl her hair, and handle men.

Henrietta looked at Gladys and whispered, "Don't you let anything bad happen to them children when I'm gone."

Then she rolled over, her back to Gladys, and closed her eyes.

Gladys slipped out of the hospital and onto a Greyhound back to Clover. That night, she called Day.

"Henrietta gonna die tonight," she told him. "She wants you to take care of them kids—I told her I'd let you know. Don't let nuthin happen to them."

Henrietta died at 12:15 a.m. on October 4, 1951.

Part Two

DEATH

12

The Storm

There was no obituary for Henrietta Lacks, but word of her death reached the Gey lab quickly. As Henrietta's body cooled in the "colored" freezer, Gey asked her doctors if they'd do an autopsy. Tissue culturists around the world had been trying to create a library of immortal cells like Henrietta's, and Gey wanted samples from as many organs in her body as possible, to see if they'd grow like HeLa. But to get those samples after her death, someone would have to ask Henrietta's husband for permission.

Though no law or code of ethics required doctors to ask permission before taking tissue from a living patient, the law made it very clear that performing an autopsy or removing tissue from the dead without permission was illegal.

The way Day remembers it, someone from Hopkins called to tell him Henrietta had died, and to ask permission for an autopsy, and Day said no. A few hours later, when Day went to Hopkins with a cousin to see Henrietta's body and sign some papers, the doctors asked again about the autopsy. They said they wanted to run tests that might help his children someday. Day's cousin said it wouldn't hurt, so eventually Day agreed and signed an autopsy permission form.

Soon Henrietta's body lay on a stainless-steel table in the cavernous basement morgue, and Gey's assistant, Mary, stood in the doorway breathing fast, feeling like she might faint. She'd never seen a dead body. Now there she was with a corpse, a stack of petri dishes, and the pathologist, Dr. Wilbur, who stood hunched over the autopsy table. Henrietta's arms were extended, as if she were reaching above her head. Mary walked toward the table, whispering to herself, *You're not going to make a fool of yourself and pass out.*

She stepped around one of Henrietta's arms and took her place beside Wilbur, her hip in Henrietta's armpit. He said hello, Mary said hello back. Then they were silent. Day wanted Henrietta to be presentable for the funeral, so he'd only given permission for a partial autopsy, which meant no incision into her chest and no removal of her limbs or head. Mary opened the dishes one by one, holding them out to collect samples as Wilbur cut them from Henrietta's body: bladder, bowel, uterus, kidney, vagina, ovary, appendix, liver, heart, lungs. After dropping each sample into a petri dish, Wilbur put bits of Henrietta's tumor-covered cervix into containers filled with formaldehyde to save them for future use.

The official cause of Henrietta's death was terminal uremia: blood poisoning from the buildup of toxins normally flushed out of the body in urine. The tumors had completely blocked her urethra, leaving her doctors unable to pass a catheter into her bladder to empty it. Tumors the size of baseballs had nearly replaced her kidneys, bladder, ovaries, and uterus. And her other organs were so covered in small white tumors it looked as if someone had filled her with pearls.

Mary stood beside Wilbur, waiting as he sewed Henrietta's abdomen closed. She wanted to run out of the morgue and back to the lab, but instead, she stared at Henrietta's arms and legs—anything to avoid looking into her lifeless eyes. Then Mary's gaze fell on Henrietta's feet, and she gasped: Henrietta's toenails were covered in chipped bright red polish.

"When I saw those toenails," Mary told me years later, "I

fainted. I thought, *Oh jeez, she's a real person.* I started imagining her sitting in her bathroom painting those toenails, and it hit me for the first time that those cells we'd been working with all this time and sending all over the world, they came from a live woman. I'd never thought of it that way."

A few days later, Henrietta's body made the long, winding train ride from Baltimore to Clover in a plain pine box, which was all Day could afford. It was raining when the local undertaker met Henrietta's coffin at the Clover depot and slid it into the back of a rusted truck. He rolled through downtown Clover, past the hardware store where Henrietta used to watch old white men play checkers, and onto Lacks Town Road, turning just before The Shack, where she'd danced only a few months earlier. As the undertaker drove into Lacks Town, cousins filed onto porches to watch Henrietta pass, their hands on hips or clutching children as they shook their heads and whispered to the Lord.

Cootie shuffled into his yard, looked straight into the falling rain, and yelled, "Sweet Jesus, let that poor woman rest, you hear me? She had enough!"

Amens echoed from a nearby porch.

A quarter-mile down the road, Gladys and Sadie sat on the broken wooden steps of the home-house, a long pink dress draped across their laps and a basket at their feet filled with makeup, curlers, red nail polish, and the two pennies they'd rest on Henrietta's eyes to keep them closed for the viewing. They watched silently as the undertaker inched through the field between the road and the house, his tires sinking into puddles of red mud.

Cliff and Fred stood in the graveyard behind the house, their overalls drenched and heavy with rain. They'd spent most of the day thrusting shovels into the rocky cemetery ground, digging a grave for Henrietta. They dug in one spot, then another, moving each time their shovels hit the coffins of unknown relatives buried with no markers.

Eventually they found an empty spot for Henrietta near her mother's tombstone.

When Cliff and Fred heard the undertaker's truck, they walked toward the home-house to help unload Henrietta. When they got her into the hallway, they opened the pine box, and Sadie began to cry. What got her most wasn't the sight of Henrietta's lifeless body, it was her toenails: Henrietta would rather have died than let her polish get all chipped like that.

"Lord," Sadie said. "Hennie must a hurt somethin worse than death."

For several days, Henrietta's corpse lay in the hallway of the home-house, doors propped open at each end to let in the cool wet breeze that would keep her body fresh. Family and neighbors waded through the field to pay respects, and all the while, the rain kept coming.

The morning of Henrietta's funeral, Day walked through the mud with Deborah, Joe, Sonny, and Lawrence. But not Elsie. She was still in Crownsville and didn't even know her mother had died.

The Lacks cousins don't remember much about the service—they figure there were some words, probably a song or two. But they all remember what happened next. As Cliff and Fred lowered Henrietta's coffin into her grave and began covering her with handfuls of dirt, the sky turned black as strap molasses. The rain fell thick and fast. Then came long rumbling thunder, screams from the babies, and a blast of wind so strong it tore the metal roof off the barn below the cemetery and sent it flying through the air above Henrietta's grave, its long metal slopes flapping like the wings of a giant silver bird. The wind caused fires that burned tobacco fields. It ripped trees from the ground, blew power lines out for miles, and tore one Lacks cousin's wooden cabin clear out of the ground, threw him from the living room into his garden, then landed on top of him, killing him instantly.

Years later, when Henrietta's cousin Peter looked back on that day, he just shook his bald head and laughed: "Hennie never was what you'd call a beatin-around-the-bush woman," he said. "We shoulda knew she was tryin to tell us somethin with that storm."

13

The HeLa Factory

Not long after Henrietta's death, planning began for a HeLa factory—a massive operation that would grow to produce trillions of HeLa cells each week. It was built for one reason: to help stop polio.

By the end of 1951 the world was in the midst of the biggest polio epidemic in history. Schools closed, parents panicked, and the public grew desperate for a vaccine. In February 1952, Jonas Salk at the University of Pittsburgh announced that he'd developed the world's first polio vaccine, but he couldn't begin offering it to children until he'd tested it on a large scale to prove it was safe and effective. And doing that would require culturing cells on an enormous, industrial scale, which no one had done before.

The National Foundation for Infantile Paralysis (NFIP)—a charity created by President Franklin Delano Roosevelt, who'd himself been paralyzed by polio—began organizing the largest field trial ever conducted to test the polio vaccine. Salk would inoculate 2 million children and the NFIP would test their blood to see if they'd become immune. But doing this would require millions of neutralization tests, which involved mixing blood serum from newly vaccinated children with live poliovirus and cells in culture. If the vaccine worked, the serum from a vaccinated child's blood would block the poliovirus and protect the cells. If it didn't work, the virus would infect the cells, causing damage scientists could see using a microscope.

The trouble was, at that point, the cells used in neutralization tests came from monkeys, which were killed in the process. This was a problem, not because of concern for animal welfare—which wasn't the issue then that it is today—but because monkeys were expensive. Doing millions of neutralization tests using monkey cells would cost millions of dollars. So the NFIP went into overdrive looking for a cultured cell that could grow on a massive scale and would be cheaper than using monkeys.

The NFIP turned to Gey and a few other cell culture experts for help, and Gey recognized the opportunity as a gold mine for the field. The NFIP's March of Dimes was bringing in an average of $50 million in donations each year, and its director wanted to give much of that money to cell culturists so they could find a way to mass-produce cells, which they'd been wanting to do for years anyway.

The timing was perfect: by chance, soon after the NFIP contacted Gey for help, he realized that Henrietta's cells grew unlike any human cells he'd seen.

Most cells in culture grew in a single layer in a clot on a glass surface, which meant they ran out of space quickly. Increasing their numbers was labor-intensive: scientists had to repeatedly scrape the cells from one tube and split them into new ones to give them more space. HeLa cells, it turned out, weren't picky—they didn't need a glass surface in order to grow. They could grow floating in a culture medium that was constantly stirred by a magnetic device, an important technique Gey developed, now called growing in suspension. This meant that HeLa cells weren't limited by space in the same way other cells were; they could simply divide until they ran out of culture medium. The bigger the vat of medium, the more the cells grew. This discovery meant that if HeLa was susceptible to poliovirus, which not all cells were, it would solve the mass-production problem and make it possible to test the vaccine without millions of monkey cells.

So in April 1952, Gey and one of his colleagues from the NFIP advisory committee—William Scherer, a young postdoctoral fellow at the University of Minnesota—tried infecting Henrietta's cells with poliovirus. Within days they found that HeLa was, in fact, *more* susceptible to the virus than any cultured cells had ever been. When they realized this, they knew they'd found exactly what the NFIP was looking for.

They also knew that, before mass-producing any cells, they'd need to find a new way to ship them. Gey's air freight shipping system worked fine for sending a few cells to colleagues here and there, but it was too expensive for shipping on a massive scale. And growing cells by the billions wouldn't help anyone if they couldn't get those cells where they needed to go. So they began experimenting.

On Memorial Day 1952, Gey gathered a handful of tubes containing HeLa cells and enough media for them to survive for a few days, and packed them into a tin lined with cork and filled with ice to prevent overheating. Then he typed up careful instructions for feeding and handling, and sent Mary to the post office to ship them to Scherer in Minnesota. Every post office in Baltimore was closed for the holiday except the main branch downtown. Mary had to take several trolleys to get there, but she made it. And so did the cells: When the package arrived in Minneapolis about four days later, Scherer put the cells in an incubator and they began to grow. It was the first time live cells had ever been successfully shipped in the mail.

In the coming months—to test different delivery methods, and make sure the cells could survive long trips in any climate—Gey and Scherer sent tubes of HeLa cells around the country by plane, train, and truck, from Minneapolis to Norwich to New York and back again. Only one tube died.

When the NFIP heard the news that HeLa was susceptible to poliovirus and could grow in large quantities for little money, it immediately contracted William Scherer to oversee development of a HeLa

Distribution Center at the Tuskegee Institute, one of the most prestigious black universities in the country. The NFIP chose the Tuskegee Institute for the project because of Charles Bynum, director of "Negro Activities" for the foundation. Bynum—a science teacher and civil rights activist who was the first black foundation executive in the country—wanted the center to be located at Tuskegee because it would provide hundreds of thousands of dollars in funding, many jobs, and training opportunities for young black scientists.

In just a few months, a staff of six black scientists and technicians built a factory at Tuskegee unlike any seen before. Its walls were lined with industrial steel autoclaves for steam sterilizing; row upon row of enormous, mechanically stirred vats of culture medium; incubators; glass culturing bottles stacked on their sides; and automatic cell dispensers—tall contraptions with long, thin metal arms that squirted HeLa cells into one test tube after another. The Tuskegee team mixed thousands of liters of Gey culture medium each week, using salts, minerals, and serum they collected from the many students, soldiers, and cotton farmers who responded to ads in the local paper seeking blood in exchange for money.

Several technicians served as a quality-control assembly line, staring through microscopes at hundreds of thousands of HeLa cultures each week, making sure the samples were alive and healthy. Others shipped them on a rigid schedule to researchers at twenty-three polio-testing centers around the country.

Eventually, the Tuskegee staff grew to thirty-five scientists and technicians, who produced twenty thousand tubes of HeLa—about 6 trillion cells—every week. It was the first-ever cell production factory, and it started with a single vial of HeLa that Gey had sent Scherer in their first shipping experiment, not long after Henrietta's death.

With those cells, scientists helped prove the Salk vaccine effective. Soon the *New York Times* would run pictures of black women hunched over microscopes examining cells, black hands holding vials of HeLa, and this headline:

UNIT AT TUSKEGEE HELPS POLIO FIGHT
Corps of Negro Scientists Has Key Role in
Evaluating of Dr. Salk's Vaccine
HELA CELLS ARE GROWN

Black scientists and technicians, many of them women, used cells from a black woman to help save the lives of millions of Americans, most of them white. And they did so on the same campus—and at the very same time—that state officials were conducting the infamous Tuskegee syphilis studies.

At first the Tuskegee Center supplied HeLa cells only to polio testing labs. But when it became clear that there was no risk of a HeLa shortage, they began sending the cells to any scientist interested in buying them, for ten dollars plus Air Express fees. If researchers wanted to figure out how cells behaved in a certain environment, or reacted to a specific chemical, or produced a certain protein, they turned to Henrietta's cells. They did that because, despite being cancerous, HeLa still shared many basic characteristics with normal cells: They produced proteins and communicated with one another like normal cells, they divided and generated energy, they expressed genes and regulated them, and they were susceptible to infections, which made them an optimal tool for synthesizing and studying any number of things in culture, including bacteria, hormones, proteins, and especially viruses.

Viruses reproduce by injecting bits of their genetic material into a living cell, essentially reprogramming the cell so it reproduces the virus instead of itself. When it came to growing viruses—as with many other things—the fact that HeLa was malignant just made it *more* useful. HeLa cells grew much faster than normal cells, and therefore produced results faster. HeLa was a workhorse: it was hardy, it was inexpensive, and it was everywhere.

And the timing was perfect. In the early fifties, scientists were just beginning to understand viruses, so as Henrietta's cells arrived in labs around the country, researchers began exposing them to viruses of all kinds—herpes, measles, mumps, fowl pox, equine encephalitis—to study how each one entered cells, reproduced, and spread.

Henrietta's cells helped launch the fledgling field of virology, but that was just the beginning. In the years following Henrietta's death, using some of the first tubes of her cells, researchers around the world made several important scientific advances in quick succession. First, a group of researchers used HeLa to develop methods for freezing cells without harming or changing them. This made it possible to send cells around the world using the already-standardized method for shipping frozen foods and frozen sperm for breeding cattle. It also meant researchers could store cells between experiments without worrying about keeping them fed and sterile. But what excited scientists most was that freezing gave them a means to suspend cells in various states of being.

Freezing a cell was like pressing a pause button: cell division, metabolism, and everything else simply stopped, then resumed after thawing as if you'd just pressed play again. Scientists could now pause cells at various intervals during an experiment so they could compare how certain cells reacted to a specific drug one week, then two, then six after exposure. They could look at identical cells at different points in time, to study how they changed with age. And by freezing cells at various points, they believed they could see the actual moment when a normal cell growing in culture became malignant, a phenomenon they called *spontaneous transformation*.

Freezing was just the first of several dramatic improvements HeLa helped bring to the field of tissue culture. One of the biggest was the standardization of the field, which, at that point, was a bit of a mess. Gey and his colleagues had been complaining that they wasted too much time just making medium and trying to keep cells alive. But more than anything, they worried that since everyone was using different media ingredients, recipes, cells, and techniques, and few knew their peers' methods, it would be difficult, if not impossible, to replicate one another's experiments. And replication is an essential part of science: a discovery isn't considered valid if others can't repeat the work and get the same result. Without standardized materials and methods, they worried that the field of tissue culture would stagnate.

Gey and several colleagues had already organized a committee to develop procedures to "simplify and standardize the technique of tissue culturing." They'd also convinced two fledgling biological supply companies—Microbiological Associates and Difco Laboratories—to begin producing and selling ingredients for culture media, and taught them the techniques necessary to do so. Those companies had just started selling media ingredients, but cell culturists still had to make the media themselves, and they all used different recipes.

Standardization of the field wasn't possible until several things happened: first, Tuskegee began mass-producing HeLa; second, a researcher named Harry Eagle at the National Institutes of Health (NIH) used HeLa to develop the first standardized culture medium that could be made by the gallon and shipped ready to use; and, third, Gey and several others used HeLa to determine which glassware and test-tube stoppers were least toxic to cells.

Only then, for the first time, could researchers around the world work with the same cells, growing in the same media, using the same equipment, all of which they could buy and have delivered to their labs. And soon they'd even be able to use the first-ever clones of human cells, something they'd been working toward for years.

Today, when we hear the word *clone,* we imagine scientists creating entire living animals—like Dolly the famous cloned sheep—using DNA from one parent. But before the cloning of whole animals, there was the cloning of individual cells—Henrietta's cells.

To understand why cellular cloning was important, you need to know two things: First, HeLa didn't grow from *one* of Henrietta's cells. It grew from a sliver of her tumor, which was a cluster of cells. Second, cells often behave differently, even if they're all from the same sample, which means some grow faster than others, some produce more poliovirus, and some are resistant to certain antibiotics. Scientists wanted to grow cellular clones—lines of cells descended from individual cells—so they could harness those unique traits. With HeLa, a group of scientists in Colorado succeeded, and soon the world of science had not only HeLa but also its hundreds, then thousands, of clones.

The early cell culture and cloning technology developed using HeLa helped lead to many later advances that required the ability to grow single cells in culture, including isolating stem cells, cloning whole animals, and in vitro fertilization. Meanwhile, as the standard human cell in most labs, HeLa was also being used in research that would advance the new field of human genetics.

Researchers had long believed that human cells contained forty-eight chromosomes, the threads of DNA inside cells that contain all of our genetic information. But chromosomes clumped together, making it impossible to get an accurate count. Then, in 1953, a geneticist in Texas accidentally mixed the wrong liquid with HeLa and a few other cells, and it turned out to be a fortunate mistake. The chromosomes inside the cells swelled and spread out, and for the first time, scientists could see each of them clearly. That accidental discovery was the first of several developments that would allow two researchers from Spain and Sweden to discover that normal human cells have forty-six chromosomes.

Once scientists knew how many chromosomes people were *supposed* to have, they could tell when a person had too many or too few, which made it possible to diagnose genetic diseases. Researchers worldwide would soon begin identifying chromosomal disorders, discovering that patients with Down syndrome had an extra chromosome number 21, patients with Klinefelter syndrome had an extra sex chromosome, and those with Turner syndrome lacked all or part of one.

With all the new developments, demand for HeLa grew, and Tuskegee wasn't big enough to keep up. The owner of Microbiological Associates—a military man named Samuel Reader—knew nothing about science, but his business partner, Monroe Vincent, was a researcher who understood the potential market for cells. Many scientists needed cells, but few had the time or ability to grow them in large enough quantities. They just wanted to buy them. So together, Reader and Vincent used HeLa cells as the springboard to launch the first industrial-scale, for-profit cell distribution center.

It started with what Reader lovingly referred to as his Cell Factory. In Bethesda, Maryland, in the middle of a wide-open warehouse that was once a Fritos factory, he built a glass-enclosed room that housed a rotating conveyor belt with hundreds of test-tube holders built into it. Outside the glass room, he had a setup much like Tuskegee's, with massive vats of culture medium, only bigger. When cells were ready for shipping, he'd sound a loud bell and all workers in the building, including the mailroom clerks, would stop what they were doing, scrub themselves at the sterilization station, grab a cap and gown, and line up at the conveyor belt. Some filled tubes, others inserted rubber stoppers, sealed tubes, or stacked them inside a walk-in incubator where they stayed until being packaged for shipping.

Microbiological Associates' biggest customers were labs like NIH, which had standing orders for millions of HeLa cells delivered on set schedules. But scientists all over the world could call in orders, pay less than fifty dollars, and Microbiological Associates would overnight them vials of HeLa cells. Reader had contracts with several major airlines, so whenever he got an order, he'd send a courier with cells to catch the next flight out, then have the cells picked up from the airport and delivered to labs by taxi. Slowly, a multibillion-dollar industry selling human biological materials was born.

Reader recruited the top minds in the field to tell him what products they needed most and show him how to make them. One of the scientists who consulted for Reader was Leonard Hayflick, arguably the most famous early cell culturist left in the field today. When I talked with him he said, "Microbiological Associates and Sam Reader were an absolute revolution in the field, and I'm not one to use the word *revolution* lightly."

As Reader's business grew, demand for cells from Tuskegee plummeted. The NFIP closed its HeLa production center because places like Microbiological Associates now supplied scientists with all the cells they needed. And soon, HeLa cells weren't the only ones being bought and sold for research—with media and equipment standardization, culturing became easier, and researchers began growing cells of all kinds. But none grew in quantities like HeLa.

As the Cold War escalated, some scientists exposed Henrietta's cells to massive doses of radiation to study how nuclear bombs destroyed cells and find ways to reverse that damage. Others put them in special centrifuges that spun so fast the pressure inside was more than 100,000 times that of gravity, to see what happened to human cells under the extreme conditions of deep-sea diving or spaceflight.

The possibilities seemed endless. At one point, a health-education director at the Young Women's Christian Association heard about tissue culture and wrote a letter to a group of researchers saying she hoped they'd be able to use it to help the YWCA's older women. "They complain that the skin and tissues of the face and neck inevitably show the wear and tear of years," she wrote. "My thought was that if you know how to keep tissue alive there must be some way of equalizing the reserve supply to the area of the throat and face."

Henrietta's cells couldn't help bring youth to women's necks, but cosmetic and pharmaceutical companies throughout the United States and Europe began using them instead of laboratory animals to test whether new products and drugs caused cellular damage. Scientists cut HeLa cells in half to show that cells could live on after their nuclei had been removed, and used them to develop methods for injecting substances into cells without destroying them. They used HeLa to test the effects of steroids, chemotherapy drugs, hormones, vitamins, and environmental stress; they infected them with tuberculosis, salmonella, and the bacterium that causes vaginitis.

At the request of the U.S. government, Gey took Henrietta's cells with him to the Far East in 1953 to study hemorrhagic fever, which was killing American troops. He also injected them into rats to see if they'd cause cancer. But mostly he tried to move on from HeLa, focusing instead on growing normal and cancerous cells from the same patient, so he could compare them to each other. But he couldn't escape the seemingly endless questions about HeLa and cell culture from other scientists. Researchers came to his lab several times each week wanting to learn his techniques, and he often traveled to labs around the world to help set up cell-culture facilities.

Many of Gey's colleagues pressured him to publish research papers so he could get credit for his work, but he always said he was too busy. At home he regularly stayed up all night to work. He applied for extensions on grants, often took months to answer letters, and at one point continued to pay a dead employee's salary for three months before anyone noticed. It took a year of nagging from Mary and Margaret for George to publish anything about growing HeLa; in the end, he wrote a short abstract for a conference, and Margaret submitted it for publication. After that, she regularly wrote and submitted his work for him.

By the mid-fifties, as more scientists began working with tissue culture, Gey became weary. He wrote to friends and colleagues saying, "Someone should coin a contemporary phrase and say, at least for the moment, 'The world has gone nuts over tissue culture and its possibilities.' I hope that some of this hullabaloo over tissue culture has at least had a few good points which have helped others . . . I wish for the most part, however, that things would settle down a bit."

Gey was annoyed by the widespread fixation on HeLa. After all, there were other cells to work with, including some he'd grown himself: A.Fi. and D-1 Re, each named after the patient it came from. He regularly offered them to scientists, but they were harder to culture, so they never took off like Henrietta's cells. Gey was relieved that companies had taken over HeLa distribution so that he didn't have to do it himself, but he didn't like the fact that HeLa was now completely out of his control.

Since the launch of the HeLa production factory at Tuskegee, Gey had been writing a steady stream of letters to other scientists, trying to restrict the way they used Henrietta's cells. At one point he wrote his longtime friend and colleague Charles Pomerat, lamenting the fact that others, including some in Pomerat's lab, were using HeLa for research Gey was "most capable" of doing himself, and in some cases had already done, but not yet published. Pomerat replied:

> With regard to your . . . disapproval for a wide
> exploration of the HeLa strain, I don't see how you can
> hope to inhibit progress in this direction since you
> released the strain so widely that it now can be purchased
> commercially. This is a little bit like requesting people
> not to work on the golden hamster! . . . I realize that it is
> the goodness of your heart that made available the HeLa
> cell and therefore why you now find that everybody
> wants to get into the act.

Pomerat suggested that Gey should have finished his own HeLa research before "releasing [HeLa] to the general public since once released it becomes general scientific property."

But Gey hadn't done that. And as soon as HeLa became "general scientific property," people started wondering about the woman behind the cells.

14

Helen Lane

So many people knew Henrietta's name, some-
one was bound to leak it. Gey had told William Scherer and his adviser
Jerome Syverton in Minneapolis, plus the people at the NFIP, who'd
probably told the team at Tuskegee. Everyone in the Gey lab knew
her name, as did Howard Jones, Richard TeLinde, and the other Hopkins doctors who'd treated her.

Sure enough, on November 2, 1953, the *Minneapolis Star* became
the first publication to name the woman behind the HeLa cells. There
was just one thing—the reporter got her name wrong. HeLa, the story
said, was "from a Baltimore woman named Henrietta Lakes."

No one knows who leaked the near-correct version of Henrietta's
name to the *Minneapolis Star.* Soon after the article ran, Gey got a letter from Jerome Syverton, saying, "I am writing to assure you that
neither Bill nor I provided the [*Minneapolis Star*] with the name of
the patient. As you know, Bill and I concur in your conviction that
the cell strain should be referred to as HeLa and that the patient's
name should not be used."

Regardless, a name was out. And two days after it was published,

Roland H. Berg, a press officer at the NFIP, sent Gey a letter saying he planned to write a more detailed article about HeLa cells for a popular magazine. Berg was "intrigued with the scientific and human interest elements in such a story," he wrote, and he wanted to learn more about it.

Gey replied saying, "I have discussed the matter with Dr. TeLinde, and he has agreed to allow this material to be presented in a popular magazine article. We must, however, withhold the name of the patient."

But Berg insisted:

> Perhaps I should describe further to you my ideas on this article, especially in view of your statement that the name of the patient must be withheld. . . . To inform [the public] you must also interest them. . . . You do not engage the attention of the reader unless your story has basic human interest elements. And the story of the HeLa cells, from what little I know of it now, has all those elements. . . .
>
> An intrinsic part of this story would be to describe how these cells, originally obtained from Henrietta Lakes, are being grown and used for the benefit of mankind. . . . In a story such as this, the name of the individual is intrinsic. As a matter of fact, if I were to proceed with the task my plan would be to interview the relatives of Mrs. Lakes. Nor would I publish the story without the full cooperation and approval of Mrs. Lakes' family. Incidentally, you may not be aware, but the identity of the patient is already a matter of public record inasmuch as newspaper reports have completely identified the individual. For example, I can refer you to the story in the *Minneapolis Star,* dated November 2, 1953.
>
> I am entirely sympathetic to your reasons for withholding the name of the patient and thus prevent a possible invasion of privacy. However, I do believe that in the

> kind of article I am projecting there would be complete
> protection of the rights of all individuals.

Berg didn't explain how releasing Henrietta's name to the public would have protected the privacy or rights of her family. In fact, doing so would have forever connected Henrietta and her family with the cells and any medical information eventually derived from their DNA. That wouldn't have protected the Lackses' privacy, but it certainly would have changed the course of their lives. They would have learned that Henrietta's cells were still alive, that they'd been taken, bought, sold, and used in research without her knowledge or theirs.

Gey forwarded the letter to TeLinde and others at Hopkins, including the head of public relations, asking how they thought he should respond.

"I see no reason why an interesting story cannot be made of it without using her name," TeLinde replied. "Since there is no reason for doing it I can see no point in running the risk of getting into trouble by disclosing it."

TeLinde didn't say what "trouble" he worried they might get into by releasing Henrietta's name. Keeping patient information confidential was emerging as a standard practice, but it wasn't law, so releasing it wasn't out of the question. In fact, he wrote Gey, "If you seriously disagree with me in this, I will be glad to talk to you."

Gey wrote to Berg saying, "An interesting story could still be built around a fictitious name." But he wasn't entirely opposed to releasing her real name. "There may still be a chance for you to win your point," he wrote. "I fully realize the importance of basic human interest elements in a story such as this and would propose therefore that you drop down to see Dr. TeLinde and myself."

Gey never told Berg that the *Minneapolis Star* article had Henrietta's name wrong, and Berg never wrote his article. But the press wasn't going away. A few months later, a reporter from *Collier's* magazine by the name of Bill Davidson contacted Gey—he was planning to write a story identical to the one Berg had proposed. This time Gey took a harder stance, perhaps because Davidson wasn't affiliated with one of Gey's major funding organizations, as Berg was. Gey agreed to be interviewed under two conditions: that he be allowed to read and approve the final article, and that the magazine not include the personal story or full name of the patient the cells came from.

The editor of the story balked. Like Berg, she wrote that "the human story behind these cells would be of great interest to the public." But Gey wouldn't budge. If she wanted him or any of his colleagues to talk with Davidson, *Collier's* would have to publish the article without the patient's name.

The editor eventually agreed, and on May 14, 1954, *Collier's* published a story about the power and promise of tissue culture. Watching HeLa cells divide on a screen, Davidson wrote, "was like a glimpse at immortality." Because of cell culture, he said, the world was "on the threshold of a hopeful new era in which cancer, mental illness and, in fact, nearly all diseases now regarded as incurable will cease to torment man." And much of that was thanks to cells from one woman, "an unsung heroine of medicine." The story said her name was Helen L., "a young woman in her thirties when she was admitted to the Johns Hopkins Hospital with an incurable cancer of the cervix." It also said Gey had grown Helen L.'s cells from a sample taken *after* her death, not before.

There's no record of where those two pieces of misinformation came from, but it's safe to assume they came from within the walls of Hopkins. As agreed, the *Collier's* editor had sent the story to Gey before publication for review. One week later she got a corrected version back from Joseph Kelly, the head of public relations at Hopkins. Kelly had rewritten the article, presumably with Gey's help, correcting several scientific errors but leaving two inaccuracies: the timing of growing the cells and the name Helen L.

Decades later, when a reporter for *Rolling Stone* asked Margaret Gey where the name Helen Lane came from, she'd say, "Oh, I don't know. It was confused by a publisher in Minneapolis. The name wasn't supposed to be revealed at all. It was just that somebody got confused."

One of Gey's colleagues told me that Gey created the pseudonym to throw journalists off the trail of Henrietta's real identity. If so, it worked. From the moment the *Collier's* article appeared until the seventies, the woman behind the HeLa cells would be known most often as Helen Lane, and sometimes as Helen Larson, but never as Henrietta Lacks. And because of that, her family had no idea her cells were alive.

15

"Too Young to Remember"

After Henrietta's funeral, cousins came from Clover and all over Turner Station to help cook for her family and care for the babies. They came and went by the dozens, bringing children and grandchildren, nieces and nephews. And one of them—no one was ever sure who—brought tuberculosis. Within weeks of Henrietta's death, Sonny, Deborah, and baby Joe—all between one and four years old—tested positive for TB.

The doctor sent Deborah home with TB pills as big as bullets, but her little brother Joe was another story. He was barely a year old, and the tuberculosis nearly killed him. Joe spent much of his second year in the hospital, coughing up blood in an isolation chamber. After that, he spent months being passed from cousin to cousin.

Because Day was working two jobs, Lawrence dropped out of school and spent most of his time taking care of his brothers and Deborah, but he wanted to get out of the house now and then to go to the pool halls. At sixteen he was too young to get in, so he lied about his age and got himself a voter's registration card saying he was eighteen. No one could prove he was lying since he'd been born on the home-house floor and didn't have a birth certificate or social security card. But his plan backfired. Because of the Korean War, Congress had just lowered the minimum age for military service to eighteen and a half, so Lawrence was drafted at sixteen. He was sent to Virginia, where he'd serve two years in a medic unit at Fort Belvoir. With Lawrence gone, someone else had to raise the Lacks children.

No one told Sonny, Deborah, or Joe what had happened to their mother, and they were afraid to ask. Back then, the rule in the house was, Do what adults say—otherwise you'll get hurt. They were to sit, hands folded, and not say a word unless someone asked them a question. As far as the children knew, their mother was there one day, gone the next. She never came back, and they got Ethel in her place.

Ethel was the woman that Sadie and Henrietta once hid from on the dance floor, the one Sadie and Margaret swore was jealous of Henrietta. They called her "that hateful woman," and when she and her husband, Galen, moved into the house, saying they were there to help with the children, Sadie and Margaret figured Ethel was trying to move in on Day. Soon, stories began spreading about Ethel sleeping with Day instead of Galen. A good handful of cousins still think Ethel moved into that house and started up with Day just to get out all the hate she had for Henrietta by torturing her children.

Henrietta's children grew up hungry. Every morning Ethel fed them each a cold biscuit that had to last them until dinner. She put latches and bolts on the refrigerator and cupboard doors to keep the children out between meals. They weren't allowed ice in their water because it made noise. If they were good, she'd sometimes give them a slice of bologna or a cold wiener, maybe pour the grease from her bacon pan onto their biscuit, or mix some water with vinegar and sugar for dessert. But she rarely thought they were good.

Lawrence came home from the military in 1953 and moved into a house of his own—he had no idea what Ethel was doing to his brothers and Deborah. As the children grew, Ethel woke them at dawn to clean the house, cook, shop, and do the laundry. In the summers she took them to Clover, where she'd send them into the fields to pick worms off tobacco leaves by hand. The tobacco juice stained their fingers and made them sick when it got in their mouths. But they grew used to it. The Lacks children had to work from sunup to sundown; they weren't allowed to take breaks, and they got no food or water until nightfall, even when the summer heat burned. Ethel would watch them from the couch or a window, and if one of them stopped working before she told them to, she'd beat them all bloody. At one point, she beat Sonny so badly with an extension cord, he ended up in the hospital. But Joe got the worst of Ethel's rage.

Sometimes she would beat Joe for no reason while he lay in bed or sat at the dinner table. She'd hit him with her fists, or whatever she had close: shoes, chairs, sticks. She made him stand in a dark basement corner on one foot, nose pressed to the wall, dirt filling his eyes. Sometimes she tied him up with rope and left him down there for hours. Other times she left him there all night. If his foot wasn't in the air when she checked on him, she'd whip his back with a belt. If he cried, she'd just whip harder. And there was nothing Sonny or Deborah could do to help him; if they said anything, Ethel just beat them all worse. But after a while it got to where the beatings didn't bother Joe. He stopped feeling pain; he felt only rage.

The police came by the house more than once to tell Day or Ethel to pull Joe off the roof, where he was lying on his stomach, shooting strangers on the sidewalk with his BB gun. When the police asked what he thought he was doing up there, Joe told them he was practicing to be a sniper when he grew up. They thought he was joking.

Joe grew into the meanest, angriest child any Lacks had ever known, and the family started saying something must have happened to his brain while he was growing inside Henrietta alongside that cancer.

In 1959, Lawrence moved into a new house with his girlfriend, Bobbette Cooper. Five years earlier she noticed Lawrence walking down the street in his uniform, and fell for him instantly. Her grandmother warned her, "Don't mess with that boy, his eyes green, his army suit green, and his car green. You can't trust him." But Bobbette didn't listen. They moved in together when Bobbette was twenty and Lawrence was twenty-four, and they had their first child that same year. They also found out that Ethel had been beating Deborah and her brothers. Bobbette insisted that the whole family move in with her and Lawrence, and she helped raise Sonny, Deborah, and Joe as if they were her own.

Deborah was ten years old. Though moving out of Ethel's house had ended the abuse for her brothers, it hadn't stopped it for her. Ethel's husband, Galen, was Deborah's biggest problem, and he found her wherever she went.

She tried to tell Day when Galen touched her in ways she didn't think he was supposed to, but Day never believed her. And Ethel just called Deborah words she'd never heard, like *bitch* and *slut.* In the car with Day driving and Ethel in the passenger seat, and everybody drinking except her, Deborah would sit in the back, pressed against the car door to get as far from Galen as she could. But he'd just slide closer. As Day drove with his arm around Ethel in front, Galen would grab Deborah in the backseat, forcing his hands under her shirt, in her pants, between her legs. After the first time he touched her, Deborah swore she'd never wear another pair of jeans with snaps instead of zippers again. But zippers didn't stop him; neither did tight belts. So Deborah would just stare out the window, praying for Day to drive faster as she pushed Galen's hands away again and again.

Then one day he called Deborah, saying, "Dale, come over here and get some money. Ethel wants you to pick her up some soda."

When Deborah got to Galen's house, she found him lying naked on the bed. She'd never seen a man's penis and didn't know what it meant for one to be erect, or why he was rubbing it. She just knew it all felt wrong.

"Ethel want a six-pack of soda," Galen told Deborah, then patted the mattress beside him. "The money's right here."

Deborah kept her eyes on the floor and ran as fast as she could, snatching the money off the bed, ducking when he grabbed for her, then running down the stairs with him chasing after her, naked and yelling, "Get back here till I finish with you, Dale! You little whore! Just wait till I tell your father!" Deborah got away, which just made him madder.

Despite the beating and the molesting, Deborah felt closer to Galen than she ever had to Day. When he wasn't hitting her, Galen showered her with attention and gifts. He bought her pretty clothes, and took her for ice cream. In those moments, Deborah pretended he was her father, and she felt like a regular little girl. But after he chased her through the house naked, it didn't seem worth it, and eventually she told Galen she didn't want any more gifts.

"I'll get you a pair of shoes," he said, then paused, rubbing her arm. "You don't have to worry about anything. I'll wear a rubber, you don't have to worry about pregnant." Deborah had never heard of a rubber, and she didn't know what pregnant was, she just knew she wanted him to leave her alone.

Deborah had started scrubbing people's floors and ironing for small amounts of money. She'd try to walk home alone after work, but Galen would usually pick her up along the way and try to touch her in the car. One day not long after her twelfth birthday, he pulled up beside Deborah and told her to get in. This time she kept walking.

Galen jammed the car into park and yelled, "You get in this damn car girl!"

Deborah refused. "Why should I get in?" she said. "I ain't doing nothing wrong, it's still daylight and I just walkin down the street."

"Your father looking for you," he snapped.

"Let him come get me then! You been doin things to my body you ain't supposed to do," she yelled. "I don't want to be nowhere with you by myself no more. Lord gave me enough sense to know that."

She turned to run but he hit her, grabbed her by the arm, threw her into the car, and kept right on having his way with her. A few weeks later, as Deborah walked home from work with a neighborhood boy named Alfred "Cheetah" Carter, Galen pulled up alongside them, yelling at her to get in the car. When Deborah refused, Galen raced up the street, tires screaming. A few minutes later he pulled up beside her again, this time with Day in the passenger seat. Galen jumped out of the car, cussing and screaming and telling her she was a whore. He grabbed Deborah by the arm, threw her in the car, and punched her hard in the face. Her father didn't say a word, just stared through the windshield.

Deborah cried the whole way home to Bobbette and Lawrence's house, blood dripping from her split eyebrow, then leapt from the car and ran through the house, straight into the closet where she hid when she was upset. She held the door closed tight. Bobbette saw Deborah run through the house crying, saw the blood on her face, and chased her to the closet. With Deborah inside sobbing, Bobbette pounded on the door saying, "Dale, what the hell is going on?"

Bobbette had been part of the family long enough to know that cousins sometimes had their way with other cousins. But she didn't know about Galen hurting Deborah, because Deborah never told anyone—she was afraid she'd get in trouble.

Bobbette pulled Deborah from the closet, grabbed her shoulders, and said, "Dale, if you don't tell me nothing, I won't know nothing. Now, I know you love Galen like he your father, but you got to tell me what's goin on."

Deborah told Bobbette that Galen had hit her, and that he sometimes talked dirty to her in the car. She said nothing about Galen touching her, because she was sure Bobbette would kill him and she worried that with Galen dead and Bobbette in jail for murdering him, she'd have lost the two people who cared for her most in the world.

Bobbette stormed over to Galen and Ethel's house, and burst in their front door screaming that if either of them touched one of those Lacks children again, she'd kill them herself.

Soon after, Deborah asked Bobbette what *pregnant* was. Bobbette told her, then grabbed Deborah's shoulders again and told her to listen good. "I know your mother and father and all the cousins all mingled together in their own way, but don't you ever do it, Dale. Cousins are not supposed to be havin sex with each other. That's uncalled for."

Deborah nodded.

"You promise me," Bobbette said. "You fight them if they try and get with you—I don't care if you have to hurt them. Don't let them touch you."

Deborah promised she wouldn't.

"You just got to go to school," Bobbette said. "Don't mess with boy cousins, and don't have babies until you're grown."

Deborah wasn't thinking about having babies anytime soon, but by the time she turned thirteen she *was* thinking about marrying that neighbor boy everyone called Cheetah, mainly because she thought Galen would have to stop touching her if she had a husband. She was also thinking she'd drop out of school.

Like her brothers, she'd always struggled in school because she couldn't hear the teacher. None of the Lacks children could hear much unless the person speaking was nearby, talking loud and slow. But they'd been taught to keep quiet with adults, so they never told their teachers how much they were missing. None of them would realize the extent of their deafness or get hearing aids until later in life.

When Deborah told Bobbette she wanted to leave school, Bobbette said, "Sit up front if you can't hear. I don't care what you do, but you get an education, cause that's your only hope."

So Deborah stayed in school. She spent summers in Clover, and as she developed, her boy cousins would grab her and try to have their way. Sometimes they'd try to drag her into a field or behind a house. Deborah fought back with fists and teeth, and soon the cousins left her alone. They'd sneer at her, tell her she was ugly, and say, "Dale mean—she born mean and she gonna stay mean." Still, three or four cousins asked Deborah to marry them and she just laughed, saying, "Man, is you crazy? That ain't no game, you know? It affects the child!"

Bobbette had told Deborah that maybe she and her siblings had hearing problems because their parents were first cousins. Deborah knew other cousins had children who were dwarves, or whose minds never developed. She wondered if that had something to do with what happened to Elsie.

Deborah didn't know she had a sister for much of her childhood. When Day finally told her, all he said was that Elsie was deaf and dumb and she'd died in an institution when she was fifteen. Deborah was devastated. She demanded to know if anyone ever tried to teach her sister sign language. No one had.

Deborah begged Lawrence to tell her about their sister, but the only thing he'd say was that she was beautiful, and that he had to take her everywhere he went so he could protect her. Deborah couldn't shake the idea that since Elsie couldn't talk, she couldn't have said no to boys like Deborah did, or tell anyone if something bad happened. Deborah hounded Lawrence to tell her anything he remembered about their sister and mother. Eventually he broke down sobbing and Deborah stopped asking.

When she was in high school, Deborah cried and lay awake at night worrying about what awful things might have happened to her mother and sister. She'd ask Day and her parents' cousins, "What in the world happened to my sister? And who was my mother? What happened to her?" Day just said the same thing again and again: "Her name was Henrietta Lacks, and she died when you was too young to remember."

16

"Spending Eternity in the Same Place"

During my first visit with Henrietta's cousin Cootie, as we sat drinking juice, he told me that no one ever talked about Henrietta. Not when she was sick, not after she died, and not now. "We didn't say words like *cancer,*" he told me, "and we don't tell stories on dead folks." At that point, he said, the family had gone so long without talking about Henrietta, it was almost like she'd never existed, except for her children and those cells.

"It sound strange," he said, "but her cells done lived longer than her memory."

If I wanted to know anything about Henrietta, he told me, I'd need to go up the road and talk to her cousin Cliff, who'd grown up with her like a brother.

When I pulled into Cliff's driveway, he figured I was a Jehovah's Witness or an insurance sales rep, since the only white people who visited him were usually one or the other. He smiled and waved just the same, saying, "How you doin?"

Cliff was in his seventies and still minding the tobacco barn behind the farmhouse his father had built decades earlier, checking the furnaces several times a day to make sure they stayed at 120 degrees. Inside Cliff's house, the electric-blue and white walls were darkened with smudges of oil and dirt. He'd blocked the stairs to the second floor with cardboard and blankets to keep warm air from going up and out through missing windows, and he'd patched holes in his ceiling, walls, and windows with newspaper and duct tape. He slept downstairs on a thin, sheetless twin bed across from the refrigerator and woodstove, next to a folding table where he'd piled so many pills, he'd forgotten what they were all for. Maybe the prostate cancer, he said. Maybe the pressure.

Cliff spent most of his time on his porch, sitting in a plaid recliner so worn down it was mostly just exposed foam and springs, waving at each car that passed. He was about six feet tall, even with several inches of slouch, his light brown skin dry and weathered like alligator, his eyes sea green at the center, with deep blue edges. Decades in shipyards and tobacco fields had left his hands coarse as burlap, his fingernails yellowed, cracked, and worn to the cuticles. As Cliff talked, he stared at the ground and twisted his arthritic fingers, one over the other like he was crossing them all for good luck. Then he untwisted them and started again.

When he heard I was writing a book about Henrietta, he got up from his recliner, pulled on a jacket, and walked over to my car, yelling, "Come on then, I'll show you where she buried!"

About a half-mile down Lacks Town Road, Cliff had me park in front of a cinder block and pressboard house that couldn't have been more than three hundred square feet inside. He jerked open a log-and-barbed-wire gate that led into a pasture and motioned for me to walk through. At the end of the pasture, hidden in the trees, stood a slave-time log cabin covered in boards with gaps wide enough to see through. Its windows had no glass and were covered by thin pieces of wood and rusted Coke signs from the fifties. The house slanted, its corners resting on piles of rocks of varying sizes that had been holding it above ground for more than two hundred years, its base high enough off the ground for a small child to crawl under.

"That there is the old home-house where Henrietta grew up!" Cliff yelled, pointing. We walked toward it through red dirt and dried leaves that cracked under our feet, the air smelling of wild roses, pine, and cows.

"Henrietta kept it nice—a real home-house. Now I can't hardly recognize it."

The floors inside were covered with straw and manure; they'd collapsed in several places under the weight of cows that now roamed free on the property. Upstairs, in the room Henrietta once shared with Day, a few remnants of life lay scattered on the floor: a tattered work boot with metal eyes but no laces, a TruAde soda bottle with a white and red label, a tiny woman's dress shoe with open toes. I wondered if it was Henrietta's.

"Could be!" Cliff said. "Sure look like her shoe."

He pointed toward what used to be the back wall, which had fallen years earlier, leaving little more than the frames of two tall windows. "This is where Henrietta slept."

She used to lie on her stomach and stare out those windows, looking at the woods and the family cemetery, a small quarter-acre clearing where a few strands of barbed wire now surrounded a scattering of tombstones. The same cows that had trampled the home-house floor had destroyed several sections of the cemetery fence. They'd left manure and hoofprints on graves, crushed flower arrangements into piles of stems, ribbon, and Styrofoam, and knocked over several tombstones, which now lay flat on the ground next to their bases.

When we got outside, Cliff shook his head and picked up fragments of a broken sign. One piece said WE LOVE, the other said MOM.

Some of the family tombstones were homemade from concrete; a few were store-bought and marble. "Them's the folks with some money," Cliff said, pointing to a marble one. Many graves were marked with index-card-sized metal plates on sticks with names and dates; the rest were unmarked.

"Used to be we'd mark them graves with a rock so we could find em," Cliff told me. "But the cemetery got cleaned out one time with a bulldozer, so that pretty much cleared those rocks on away." There were so many people buried in the Lacks cemetery now, he said, they'd run out of room decades ago and started piling graves on top of each other.

He pointed at an indentation in the ground with no marker beside it. "This was a good friend of mine," he said. Then he started pointing around the graveyard to other body-sized indentations in the dirt. "See that sunk in right there . . . and that sunk right there . . . and there . . . Them's all unmarked graves. They sink after a time when the dirt settle around the bodies." Occasionally he'd point to a small plain rock poking through the earth and say it was a cousin or an aunt.

"That there's Henrietta's mother," he said, pointing to a lone tombstone near the cemetery's edge, surrounded by trees and wild roses. It was several feet tall, its front worn rough and browned from age and weather. The inscription said this:

ELIZA

WIFE OF J.R.

PLEASANT

JUL 12, 1888

OCTOBER 28, 1924

GONE BUT NOT FORGOTTEN

Until I read those dates, I hadn't done the math: Henrietta was barely four years old when she lost her mother, about the same age Sonny was when Henrietta died.

"Henrietta used to come talk to her mother, took real good care of her grave. Now Henrietta somewhere in here with her," Cliff said, waving his arms toward the clearing between Eliza's stone and the next tree a good fifteen feet away. "Never did get a marker, so I couldn't tell you exactly where she at, but the immediate family be buried next to each other. So she probably round in here somewhere."

He pointed to three body-sized indentations in the clearing and said, "Any one of those could be Henrietta."

We stood in silence as Cliff kicked at the dirt with his toe.

"I don't know what happened on that deal with them cells from Henrietta," he said eventually. "Don't nobody say anything about it round here. I just knowed she had something rare, cause she been dead a pretty good while, but her cells still living, and that's amazing." He kicked at the ground. "I heard they did a lot of research and some of her cells have develop a lot of curing other diseases. It's a miracle, that's all I can say."

Then suddenly he yelled at the ground, as if he was talking directly to Henrietta. "They named them HeLa! And they still living!" He kicked at the dirt again.

A few minutes later, seemingly out of nowhere, he pointed to the dirt and said, "You know, white folks and black folks all buried over top of each other in here. I guess old white granddaddy and his brothers was buried in here too. Really no tellin who in this ground now." Only thing he knew for sure, he said, was that there was something beautiful about the idea of slave-owning white Lackses being buried under their black kin.

"They spending eternity in the same place," he told me, laughing. "They must've worked out their problems by now!"

Henrietta's great-great-grandmother was a slave named Mourning. A white man named John Smith Pleasants inherited Mourning and her husband, George, from his father, one of the first slaveholders in Clover. Pleasants' father came from a family of Quakers, and one of his distant relatives had been the first to fight successfully to free his own slaves through the Virginia courts. But Pleasants hadn't carried on the family's antislavery tradition.

Mourning and George were enslaved on a tobacco plantation in Clover. Their son, Henrietta's paternal great-grandfather Edmund, took his owner's last name, which lost the s to become Pleasant. He was eventually freed from slavery at the age of forty, only to be committed later to an asylum for dementia. But before he was freed, he fathered many children, all of them born into slavery, including a daughter named Henrietta Pleasant—the great-aunt of Henrietta Lacks.

On the other side of Henrietta's family, her maternal great-grandfather was a white man named Albert Lacks, who'd inherited part of the Lacks Plantation in 1885, when his father divided his land among his three white sons: Winston, Benjamin, and Albert.

Winston Lacks was a burly man with a beard that grew to his belly—he drank almost every night in a saloon hidden in the basement beneath the general store. When Winston got drunk and started fighting, the locals knew it was time for the soberest man to ride and get Fannie. There are no records of Fannie's life, but she was most likely born a slave on the Lacks property, and like most Lacks slaves who stayed on the plantation as sharecroppers, she never left. She often rode beside Winston in his wagon, and when he got drunk, she'd march into the saloon, snatch him off the barstool by his long beard, and drag him home.

The other brothers, Albert and Benjamin, led more private lives and left behind little history aside from their wills and land deeds. Most of the black Lackses I talked to over the years referred to Benjamin Lacks as "old white granddaddy," though some still called him "Massuh Ben," as their parents had.

When Albert died on February 26, 1889, slavery had been abolished, but few black people owned land of their own. Albert's will left land to five "colored" heirs, most of it in ten-acre chunks, and one of those heirs was Henrietta and Day's grandfather, Tommy Lacks. Albert's will said nothing of his relation to his heirs, but folks in Lacks Town knew they were children he'd had with a former slave named Maria.

After Albert's death, his brother Benjamin sued to take some of that land away from Albert's black heirs, saying that since it was his father's land originally, he had the right to choose whichever plot he wanted. The court agreed and divided the original Lacks plantation into two plots "of equal value." The lower section—on the river—went to Benjamin Lacks; the upper plot—now known as Lacks Town— went to the black Lackses.

Sixteen years after the court case, when Benjamin Lacks dictated his own will days before his death, he gave small plots of land to each of his sisters, then divided the remaining 124 acres and his horses between seven "colored" heirs of his own, including his nephew Tommy Lacks. There's no record of Benjamin or Albert Lacks marrying or having any white children, and as with Albert, there's no record that the black children in Benjamin's will were his own. But he called them his "nigger children," and according to black Lacks oral history, everyone living on the land in Clover that was once the Lacks Plantation descended from those two white brothers and their black mistresses who were once slaves.

When I arrived in Clover, race was still ever-present. Roseland was "the nice *colored* fellow" who ran Rosie's before it shut down; Bobcat was "the *white* man" who ran the mini-mart; Henrietta went to St. Matthew's, "the *colored* church." One of the first things Cootie said when I met him was, "You don't act strange around me cause I'm black. You're not from around here."

Everyone I talked to swore race relations were never bad in Clover. But they also said Lacks Town was only about twelve miles from the local Lynch Tree, and that the Ku Klux Klan held meetings on a school baseball field less than ten miles from Clover's Main Street until well into the 1980s.

Standing in the cemetery, Cliff told me, "The white Lackses know their kin all buried in here with ours cause they family. They know it, but they'll never admit it. They just say, 'Them Black Lackses, they ain't kin!' "

When I went to visit Carlton and Ruby Lacks, the oldest white Lackses in Clover, they smiled and chitchatted as they led me from their front door into a living room filled with pastel-blue overstuffed chairs and Confederate flags—one in each ashtray, several on the coffee table, and a full-sized one on a stand in the corner. Carlton and Ruby were distant cousins before they became husband and wife. They were both related to Robin Lacks, the father of Albert, Ben, and Winston Lacks, which meant they were also Henrietta and Day's distant cousins.

Carlton and Ruby had been married for decades and had more children, grandchildren, and great-grandchildren than they could count. All they knew for sure was that there were more than one hundred of them. Carlton was a frail man in his late eighties, with skin so pale it looked almost translucent. Tufts of hair like overgrown cotton sprouted from his head, brow, ears, and nostrils as he sat in his easy chair, mumbling about his years working the bank at a tobacco warehouse.

"I wrote out the checks," he said, mostly to himself. "I was the tobacco king."

Ruby was in her late eighties too, with a sharp mind that seemed decades younger than her frail body. She talked right over Carlton, telling me about their relatives who'd farmed the Lacks Plantation, and their relation to Ben and Albert Lacks. When I mentioned that Henrietta came from Lacks Town, Ruby straightened in her chair.

"Well, that was colored!" she snapped. "I don't know what you talking about. You're not talking about coloreds are you?"

I told her I wanted to learn about both the white and black Lackses.

"Well, we never did know each other," she said. "The white and the black didn't mix then, not like they do now, which I can't say I like because I don't think it's for the best." She paused and shook her head. "Mixing them like that, during school and church and everything, they end up white and black get together and marry and all . . . I just can't see the sense in it."

When I asked how she and Carlton were related to the black Lackses, they looked at each other from across the coffee table like I'd asked if they were born on Mars.

"My daddy's uncle kept a lot of the colored Lackses as slaves," Ruby said. "That must be where they got their name. Evidently they took it when they left the plantation. That's the only thing I can figure."

Later, I asked Henrietta's sister Gladys what she thought of their theory. Though she'd lived about a mile from Carlton and Ruby Lacks most of her ninety years, Gladys said she'd never heard of them.

"Black and white Lacks is kin," Gladys said, "but we don't mix."

She pointed under the couch where I was sitting.

"Get Lillian's letter," she said to her son Gary.

As far as Gladys knew, all of Henrietta's other siblings were dead, except maybe Lillian, the youngest. The last anyone had heard from Lillian was a letter she'd sent sometime in the eighties, which Gladys kept in a shoebox under the couch. In it, Lillian wrote, "I heard daddy died in a fire," and she asked if it was true. It was: He'd died in 1969, two decades before she sent that letter. But what Lillian really wanted to know was who'd been talking to people about her life. She'd won the lottery, she said, and she believed someone was trying to kill her because white folks had been coming around asking questions about her life in Clover and her family, especially Henrietta. "They knew things I didn't even know," she wrote. "I don't think anybody should talk about other people." No one in the family had heard from her since.

"Lillian converted to Puerto Rican," Gladys said, holding the letter to her chest.

I looked at Gary, who sat beside her.

"Lillian's skin was real light, even lighter than mom's," Gary explained. "She married a Puerto Rican somewhere in New York. Since she could pass, she disowned her blackness—converted to Puerto Rican because she didn't want to be black no more."

17

Illegal,
Immoral, and
Deplorable

As HeLa grew like crabgrass in laboratories around the world, a virologist named Chester Southam had a frightening thought: What if Henrietta's cancer cells could infect the scientists working on them? Gey and several others had already shown that some rats grew tumors when injected with live HeLa. Why not humans?

Researchers were breathing in the air around HeLa cells, touching them and transferring them from vial to vial, even eating lunch at lab tables beside them. One had used them to grow a vaccine for a common-cold-like virus, which he'd injected—along with bits of HeLa—into more than four hundred people. Yet no one knew whether a person could actually catch cancer from HeLa or other cancer cells.

"There is the possible danger," Southam wrote, "of initiating neoplastic disease by accidental inoculation during laboratory investigation, or by injection with such cells or cell products if they should be used for production of virus vaccine."

Southam was a well-respected cancer researcher and chief of virology at Sloan-Kettering Institute for Cancer Research. He and many other scientists believed that cancer was caused by either a virus or an immune system deficiency, so Southam decided to use HeLa to test those theories.

In February 1954, Southam loaded a syringe with saline solution mixed with HeLa. He slid the needle into the forearm of a woman who'd recently been hospitalized for leukemia, then pushed the plunger, injecting about five million of Henrietta's cells into her arm. Using a second needle, Southam tattooed a tiny speck of India ink next to the small bump that formed at the HeLa injection site. That way, he'd know where to look when he reexamined the woman days, weeks, and months later, to see if Henrietta's cancer was growing on her arm. He repeated this process with about a dozen other cancer patients. He told them he was testing their immune systems; he said nothing about injecting them with someone else's malignant cells.

Within hours, the patients' forearms grew red and swollen. Five to ten days later, hard nodules began growing at the injection sites. Southam removed some of the nodules to verify that they were cancerous, but he left several to see if the patients' immune systems would reject them or the cancer would spread. Within two weeks, some of the nodules had grown to two centimeters—about the size of Henrietta's tumor when she went in for her radium treatments.

Southam eventually removed most of the HeLa tumors, and those he didn't remove vanished on their own in a few months. But in four patients, the nodules grew back. He removed them, but they returned again and again. In one patient, Henrietta's cancer cells metastasized to her lymph nodes.

Since those patients had all had cancer to begin with, Southam wanted to see how healthy people reacted to the injections, for comparison's sake. So, in May 1956, he placed an ad in the Ohio State Penitentiary newsletter: *Physician seeks 25 volunteers for cancer research*. A few days later he had ninety-six volunteers, which quickly increased to 150.

He chose the Ohio prison because its inmates had cooperated in several other studies without resistance, including one in which they'd been infected with a potentially deadly disease called tularemia. Research on inmates would come under scrutiny and start being heavily regulated about fifteen years later, because they'd be considered a vulnerable population unable to give informed consent. But at the time, prisoners nationwide were being used for research of all kinds—from testing chemical warfare agents to determining how X-raying testicles affected sperm count.

Southam began injecting prisoners in June 1956 using HeLa cells that his colleague, Alice Moore, carried from New York to Ohio in a handbag. Sixty-five prisoners—murderers, embezzlers, robbers, and forgers—lined up on wooden benches for their injections. Some wore white hospital garb; others came off work gangs wearing blue dungarees.

Soon tumors grew on the prisoners' arms just as they'd grown in the cancer patients. The press ran story after story about the brave men at the Ohio Penitentiary, praising them as "the first healthy human beings ever to agree to such rigorous cancer experiments." They quoted one man saying, "I'd be lying if I said I wasn't worried. You lie there on your bunk knowing you've got cancer in your arm. . . . Boy, what you think about!"

Again and again reporters asked, "Why did you volunteer for this test?"

The prisoners' replies were like a refrain: "I done a girl a great injustice, and I think it'll pay back a little bit what I did to her."

"I believe the wrong that I have done, in the eyes of society, this might make a right on it."

Southam gave multiple cancer cell injections to each prisoner, and unlike the terminally ill patients, those men fought off the cancer completely. And with each new injection, their bodies responded faster, which seemed to indicate that the cells were increasing the inmates' immunity to cancer. When Southam reported his results, the press hailed them as a tremendous breakthrough that could someday lead to a cancer vaccine.

In the coming years, Southam injected HeLa and other living cancer cells into more than six hundred people for his research, about half of them cancer patients. He also began injecting them into every gynecologic surgery patient who came to Sloan-Kettering's Memorial Hospital or its James Ewing Hospital. If he explained anything, he simply said he was testing them for cancer. And he believed he was: Since people with cancer seemed to reject the cells more slowly than healthy people did, Southam thought that by timing the rejection rate, he might be able to find undiagnosed cases of cancer.

In a statement he'd later repeat again and again during hearings about his research, Southam wrote, "It is, of course, inconsequential whether these are cancer cells or not, since they are foreign to the recipient and hence are rejected. The only drawback to the use of cancer cells is the phobia and ignorance that surrounds the word *cancer*."

Because of that "phobia and ignorance," Southam wrote, he didn't tell patients the cells were cancerous because he didn't want to cause any unnecessary fear. As he would say, "To use the dreaded word 'cancer' in connection with any clinical procedure on an ill person is potentially deleterious to that patient's well-being, because it may suggest to him (rightly or wrongly) that his diagnosis is cancer or that his prognosis is poor. . . . To withhold such emotionally disturbing but medically nonpertinent details . . . is in the best tradition of responsible clinical practice."

But Southam wasn't their doctor, and he wasn't withholding upsetting health information. The deception was for his benefit—he was withholding information because patients might have refused to participate in his study if they'd known what he was injecting. And Southam probably would have continued doing this for years had he not made an arrangement on July 5, 1963, with Emanuel Mandel, director of medicine at the Jewish Chronic Disease Hospital in Brooklyn, to use the hospital's patients for his research.

The plan was that Mandel would have doctors on his staff inject twenty-two JCDH patients with cancer cells for Southam. But when he instructed his staff to give the injections without telling patients they contained cancer cells, three young Jewish doctors refused, saying they wouldn't conduct research on patients without their consent.

All three knew about the research Nazis had done on Jewish prisoners. They also knew about the famous Nuremberg Trials.

Sixteen years earlier, on August 20, 1947, a U.S.-led war tribunal in Nuremberg, Germany, had sentenced seven Nazi doctors to death by hanging. Their crime was conducting unthinkable research on Jews without consent—sewing siblings together to create Siamese twins, dissecting people alive to study organ function.

The tribunal set forth a ten-point code of ethics now known as the Nuremberg Code, which was to govern all human experimentation worldwide. The first line in that code says, "The voluntary consent of the human subject is absolutely essential." The idea was revolutionary. The Hippocratic Oath, written in the fourth century BC, didn't require patient consent. And though the American Medical Association had issued rules protecting laboratory animals in 1910, no such rules existed for humans until Nuremberg.

But the Nuremberg Code—like other codes that would come after it—wasn't law. It was, essentially, a list of recommendations. It wasn't routinely taught in medical schools, and many American researchers—including Southam—claimed not to know it existed. Those who did know about it often thought of it as "the Nazi code," something that applied to barbarians and dictators, not to American doctors.

When Southam began injecting people with HeLa cells in 1954, there was no formal research oversight in the United States. Since the turn of the century, politicians had been introducing state and federal laws with hopes of regulating human experimentation, but physicians and researchers always protested. The bills were repeatedly voted down for fear of interfering with the progress of science, even though other countries—including, ironically, Prussia—had enacted regulations governing human research as early as 1891.

In the United States, the only way to enforce research ethics was in the civil courts. There, lawyers could use the Nuremberg Code to establish whether a scientist was acting within the ethical boundaries of the profession. But taking a researcher to court required money, know-how, and the knowledge that you were being used for research in the first place.

The term *informed consent* first appeared in court documents in 1957, in a civil court ruling on the case of a patient named Martin Salgo. He went under anesthesia for what he thought was a routine procedure and woke up permanently paralyzed from the waist down. The doctor hadn't told him the procedure carried any risks at all. The judge ruled against the doctor, saying, "A physician violates his duty to his patient and subjects himself to liability if he withholds any facts which are necessary to form the basis of an intelligent consent by the patient to the proposed treatment." He wrote that there needed to be "full disclosure of facts necessary to an informed consent."

Informed consent focused on what doctors were required to tell their patients; there was little mention of how it might apply to research like Southam's, in which subjects weren't the researcher's patients. And it would be decades before anyone thought to ask whether informed consent should apply in cases like Henrietta's, where scientists conduct research on tissues no longer attached to a person's body.

But to the three doctors who refused to help with Southam's research, injecting cancer cells into a person without consent was a clear violation of basic human rights and the Nuremberg Code. Mandel didn't see it that way. He had a resident give the injections in their place, and on August 27, 1963, the three doctors wrote a resignation letter citing unethical research practices. They sent it to Mandel and at least one reporter. When Mandel got the letter, he called a meeting with one of the doctors, and accused them of being overly sensitive because of their Jewish ancestry.

One member of the hospital's board of directors, a lawyer named William Hyman, didn't think they were being overly sensitive. When he heard about the doctors' resignation, he asked to see the records of patients in the study. But his request was denied. Meanwhile, just days after the doctors resigned, the *New York Times* ran a tiny news item deep in the paper under the headline SWEDEN PENALIZES CANCER SPECIALIST, about a cancer researcher named Bertil Björklund. He'd been giving himself and patients intravenous injections of vaccines made from HeLa cells, which he'd gotten from George Gey's lab in such enormous quantities, they joked that instead of injecting them, Björklund could just fill a pool with HeLa—or maybe even a lake—and swim around in it for immunity. Björklund's HeLa injections got him expelled from his laboratory, and Hyman hoped for similar results with Southam. So, in December 1963 he sued the hospital for access to medical records related to the study.

Hyman compared Southam's study to Nazi research and got affidavits from the three doctors who'd resigned—they described Southam's research using words like *illegal, immoral,* and *deplorable.* Hyman also got an affidavit from a fourth doctor explaining that the patients in the study wouldn't have been capable of giving informed consent even if Southam had asked: one had advanced Parkinson's disease and couldn't talk, others spoke only Yiddish, one had multiple sclerosis and "depressive psychosis." Regardless, Hyman wrote, "I was informed that consent was not necessary . . . that it was unlikely that Jewish patients would agree to live cancer cell injections."

That caught the media's attention. The hospital called the suit "misleading and fallacious." But newspapers and magazines ran headlines saying:

PATIENTS INJECTED WITH CELLS NOT TOLD THEY WERE CANCER . . . SCIENTIFIC EXPERTS CONDEMN ETHICS OF CANCER INJECTION

They said the Nuremberg Code didn't seem to apply in the United States, and that there were no laws protecting research subjects. *Science* magazine called it "the hottest public debate on medical ethics since the Nuremberg trials," and said, "The situation at present appears rather perilous for everyone." A reporter from *Science* asked

Southam why, if the injections were as safe as he swore they were, he didn't inject himself.

"Let's face it," Southam responded, "there are relatively few skilled cancer researchers, and it seemed stupid to take even the little risk."

Patients who'd unknowingly been injected with cancer cells by Southam read the articles and began contacting reporters. New York State Attorney General Louis Lefkowitz learned about Southam's research through the media as well, and immediately launched his own investigation. In a scathing five-page document filled with exclamation points, he accused Southam and Mandel of fraud and unprofessional conduct, and demanded that the Board of Regents of the University of the State of New York revoke their medical licenses. Lefkowitz wrote, "Every human being has an inalienable right to determine what shall be done with his own body. These patients then had a right to know . . . the contents of the syringe: and if this knowledge was to cause fear and anxiety or make them frightened, they had a right to be fearful and frightened and thus say NO to the experiment."

Many doctors testified before the Board of Regents and in the media on Southam's behalf, saying they'd been conducting similar research for decades. They argued that it was unnecessary to disclose all information to research subjects or get consent in all cases, and that Southam's behavior was considered ethical in the field. Southam's lawyers argued, "If the whole profession is doing it, how can you call it 'unprofessional conduct'?"

This rattled the Board of Regents. On June 10, 1965, its Medical Grievance Committee found Southam and Mandel guilty of "fraud or deceit and unprofessional conduct in the practice of medicine" and recommended that their medical licenses be suspended for one year. The Board wrote, "There is evidenced in the record in this proceeding an attitude on the part of some physicians that they can go ahead and do anything . . . and that the patient's consent is an empty formality. With this we cannot agree."

Their decision called for more specific guidelines in clinical research, saying, "We trust that this measure of discipline will serve as a stern warning that zeal for research must not be carried to the point where it violates the basic rights and immunities of a human person."

The suspensions of Southam's and Mandel's licenses were stayed, leaving them both on one-year probation instead. And the case seemed to have little impact on Southam's professional standing: soon after the end of his probationary period, Southam was elected president of the American Association for Cancer Research. But his case brought about one of the largest research oversight changes in the history of experimentation on humans.

Before the Board of Regents announced its decision, the negative press about Southam's work had gotten the attention of the NIH, which funded his research and required its investigators to get consent for all studies involving humans. In response to the Southam situation, the NIH investigated all their grantee institutions and found that only nine out of fifty-two had any policy in place to protect the rights of research subjects. Only sixteen used consent forms. The NIH concluded: "In the setting in which the patient is involved in an experimental effort, the judgment of the investigator is not sufficient as a basis for reaching a conclusion concerning the ethical and moral set of questions in that relationship."

As a result of its investigation, the NIH said that to qualify for funding, all proposals for research on human subjects had to be approved by review boards—independent bodies made up of professionals and laypeople of diverse races, classes, and backgrounds—to ensure that they met the NIH's ethics requirements, including detailed informed consent.

Scientists said medical research was doomed. In a letter to the editor of *Science,* one of them warned, "When we are prevented from attempting seemingly innocuous studies of cancer behavior in humans . . . we may mark 1966 as the year in which all medical progress ceased."

Later that year, a Harvard anesthesiologist named Henry Beecher published a study in the *New England Journal of Medicine* showing that Southam's research was only one of hundreds of similarly unethical studies. Beecher published a detailed list of the twenty-two worst offenders, including researchers who'd injected children with hepatitis and others who'd poisoned patients under anesthesia using carbon dioxide. Southam's study was included as example number 17.

Despite scientists' fears, the ethical crackdown didn't slow scientific progress. In fact, research flourished. And much of it involved HeLa.

18

"Strangest Hybrid"

By the 1960s, scientists joked that HeLa cells were so robust that they could probably survive in sink drains or on doorknobs. They were everywhere. The general public could grow HeLa at home using instructions from a *Scientific American* do-it-yourself article, and both Russian and American scientists had managed to grow HeLa in space.

Henrietta's cells went up in the second satellite ever in orbit, which was launched by the Russian space program in 1960, and almost immediately afterward, NASA shot several vials of HeLa into space in the *Discoverer XVIII* satellite. Researchers knew from simulated zero-gravity studies using animals that space travel could cause cardiovascular changes, degradation of bone and muscle, and a loss of red blood cells. They also knew radiation levels were higher beyond the ozone layer. But they didn't know what effects any of this would have on humans: Would it cause cellular changes, or even cell death?

When the first humans went into orbit, Henrietta's cells went with them so researchers could study the effects of space travel, as well as the nutritional needs of cells in space, and how cancerous and noncancerous cells responded differently to zero gravity. What they found was disturbing: in mission after mission, noncancerous cells grew normally in orbit, but HeLa became more powerful, dividing faster with each trip.

And HeLa cells weren't the only ones behaving strangely. Since the start of the decade, researchers had been noticing two new things about all cultured cells. First, it seemed that all normal cells growing in culture eventually died or underwent spontaneous transformation and became cancerous. This phenomenon was exciting for researchers trying to understand the mechanisms of cancer, because it suggested that they might be able to study the moment a normal cell becomes malignant. But it was disturbing for those trying to use cell culture to develop medical therapies.

George Hyatt, a Navy doctor working with the National Cancer Institute, had experienced this phenomenon firsthand. He'd cultured human skin cells for treating badly burned soldiers, then created a wound on a young volunteer officer's arm and smeared the cells across it, hoping they'd grow to form a new layer of skin. If it worked, it might mean doctors could use skin-cell transplants to treat wounds in the field. The cells did grow, but when Hyatt biopsied them a few weeks later, they were all cancerous. He panicked, removed the cells, and hadn't tried transplanting skin cells since.

The other unusual thing scientists had noticed about cells growing in culture was that once they transformed and became cancerous, they all behaved alike—dividing identically and producing exactly the same proteins and enzymes, even though they'd all produced different ones before becoming malignant. Lewis Coriell, a renowned cell culturist, thought he might have an explanation. He published a paper suggesting that perhaps "transformed" cells behaved the same not because they'd become cancerous, but because they'd been contaminated by something—most likely a virus or bacterium—that made them behave similarly. Almost as an aside, he pointed out one possibility that other researchers hadn't considered: all transformed cells seemed to behave identically to HeLa, he wrote, which could mean that HeLa was the contaminant.

Soon after his paper was published, Coriell and a few other top tissue culturists called an urgent meeting to talk about the state of their field, which they worried was becoming a disaster. They'd mastered the techniques of cell culture and simplified them to such a degree that, as one researcher put it, they'd "made it possible for even the rank amateur to grow a few cultures."

In recent years, using tissue samples from themselves, their families, and their patients, scientists had grown cells of all kinds—prostate cancer, appendix, foreskin, even bits of human cornea—often with surprising ease. Researchers were using that growing library of cells to make historic discoveries: that cigarettes caused lung cancer; how X-rays and certain chemicals transformed normal cells into malignant ones; why normal cells stopped growing and cancer cells didn't. And the National Cancer Institute was using various cells, including HeLa, to screen more than thirty thousand chemicals and plant extracts, which would yield several of today's most widely used and effective chemotherapy drugs, including Vincristine and Taxol.

Despite the importance of this research, many scientists seemed cavalier about their cultures. Few kept clear records of which cells grew from which donors, and many mislabeled their cultures, if they labeled them at all. For scientists doing research that *wasn't* cell-specific, like investigating the effects of radiation on DNA, not knowing what kind of cell they were working on might not affect the outcome of their research. But if cells were contaminated or mislabeled in research that *was* cell-specific—as much research was—the results would be worthless. Regardless, the culturists who called the meeting said, precision was essential in science, and researchers should know what cells they were using, and whether they were contaminated.

According to Robert Stevenson, one of the scientists involved in the meeting, their goal was to keep the field from "degenerating into complete chaos." The group encouraged researchers to use protective measures, like working under hoods with suction that pulled air and potential contaminants into a filtration system. And they recommended that the NIH establish a reference collection of cells: a central bank where all cultures would be tested, cataloged, and stored under maximum security, using state-of-the-art sterile techniques. The NIH agreed, and formed a Cell Culture Collection Committee made up of tissue culturists, including William Scherer, Lew Coriell, and Robert Stevenson. Their mission was to establish a nonprofit federal cell bank at the American Type Culture Collection (ATCC), which had been distributing and monitoring the purity of bacteria, fungi, yeast, and viruses since 1925, but never cultured cells.

The scientists on the Collection Committee set out to create the Fort Knox of pure, uncontaminated cell culture. They transported cultures in locked suitcases and developed a list of criteria all cells had to meet before being banked: each had to be tested for any possible contamination, and they all had to come directly from the original source.

Cell number one in the ATCC's collection was the L-cell, the original immortal mouse cell line grown by Wilton Earle. For cell number two, the committee contacted Gey asking for a sample from the original HeLa culture. But in his initial excitement, Gey had given all of the original HeLa cells to other researchers and kept none for himself. He eventually tracked some down in the lab of William Scherer, who'd used some of the original HeLa sample in their polio research.

Initially the committee could only test samples for viral and bacterial contamination, but soon a few of its members developed a test for cross-species contamination, so they could determine whether cultures labeled as being from one animal type were actually from another. They quickly found that of ten cell lines thought to be from nine different species—including dog, pig, and duck—all but one were actually from primates. They promptly relabeled those cultures, and it seemed they'd gotten the situation under control without attracting any bad publicity.

The media, it turned out, was far more interested in a bit of HeLa-related news that was almost as sensational as Alexis Carrel's immortal chicken heart. And it all started with cell sex.

In 1960, French researchers had discovered that when cells were infected with certain viruses in culture, they clumped together and sometimes fused. When they fused, the genetic material from the two cells combined, as with sperm meeting egg. The technical name for this was *somatic cell fusion,* but some researchers called it "cell sex." It was different from sperm-and-egg sex in several important ways: somatic cells were cells of the body, like skin cells, and their union produced offspring every few hours. Perhaps most important, cell sex was entirely controlled by researchers.

Genetically speaking, humans are terrible research subjects. We're genetically promiscuous—we mate with anyone we choose—and we don't take kindly to scientists telling us who to reproduce with. Plus, unlike plants and mice, it takes us decades to produce enough offspring to give scientists much meaningful data. Since the mid-1800s, scientists had studied genes by breeding plants and animals in specific ways—a smooth pea with a wrinkled one, a brown mouse with a white one—then breeding their offspring to see how genetic traits passed from one generation to the next. But they couldn't study human genetics the same way. Cell sex solved that problem, because it meant researchers could combine cells with any traits they wanted and study how those traits were passed along.

In 1965 two British scientists, Henry Harris and John Watkins, took cell sex an important step further. They fused HeLa cells with mouse cells and created the first human-animal hybrids—cells that contained equal amounts of DNA from Henrietta and a mouse. By doing this, they helped make it possible to study what genes do, and how they work.

In addition to the HeLa-mouse hybrid, Harris fused HeLa with chicken cells that had lost their ability to reproduce. His hunch was that when those deactivated chicken cells fused with HeLa, something inside HeLa would essentially turn the chicken cell back on. He was right. He didn't know how it worked yet, but his discovery showed that something in cells regulated genes. And if scientists could figure out how to turn disease genes off, they might be able to create a form of gene therapy.

Soon after Harris's HeLa-chicken study, a pair of researchers at New York University discovered that human-mouse hybrids lost their human chromosomes over time, leaving only the mouse chromosomes. This allowed scientists to begin mapping human genes to specific chromosomes by tracking the order in which genetic traits vanished. If a chromosome disappeared and production of a certain enzyme stopped, researchers knew the gene for that enzyme must be on the most recently vanished chromosome.

Scientists in laboratories throughout North America and Europe began fusing cells and using them to map genetic traits to specific chromosomes, creating a precursor to the human genome map we have today. They used hybrids to create the first monoclonal antibodies, special proteins later used to create cancer therapies like Herceptin, and to identify the blood groups that increased the safety of transfusions. They also used them to study the role of immunity in organ transplantation. Hybrids proved it was possible for DNA from two unrelated individuals, even of different species, to survive together *inside* cells without one rejecting the other, which meant the mechanism for rejecting transplanted organs had to be *outside* cells.

Scientists were ecstatic about hybrids, but throughout the United States and Britain, the public panicked as the media published one sensational headline after the next:

MAN-ANIMAL CELLS ARE BRED IN LAB . . . THE NEXT STEP COULD BE TREE MEN . . . SCIENTISTS CREATE MONSTERS

The Times of London called the HeLa-mouse cells the "strangest hybrid form of life ever seen in the lab—or out of it." A *Washington Post* editorial said, "We cannot afford any artificially induced mouse-men."

It called the research "horrendous" and said the researchers should leave humans alone and "go back to their yeasts and fungi." One article ran with an image of a half-human, half-mouse creature with a long, scaly tail; another ran with a cartoon of a hippopotamus-woman reading the newspaper at a bus stop. The British press called the HeLa hybrids an "assault on life," and portrayed Harris as a mad scientist. And Harris didn't help the situation: he caused near-pandemonium when he appeared in a BBC documentary saying that the eggs of man and ape could now be joined to create a "mape."

Harris and Watkins wrote letters to editors complaining they'd been quoted out of context, their story sensationalized to "distort, misrepresent and terrify." They assured the public that they were just creating cells, not "trying to produce centaurs." But it didn't help. A public survey about their research was overwhelmingly negative, calling it pointless and dangerous, an example of "men trying to be gods." And the PR problem for cell culture was only going to get worse from there.

19

"The Most Critical Time on This Earth Is Now"

When Deborah was a junior in high school, at the age of sixteen, she got pregnant with her first child. Bobbette cried when she found out. Deborah stopped going to school and Bobbette said, "Don't get too comfortable cause you're goin to graduate." Deborah yelled right back, saying she couldn't go to school all big and pregnant.

"That don't matter," Bobbette said, "you're goin to that special girls school where all the pregnant girls have big bellies just like you."

Deborah refused, but Bobbette filled out the application for her and dragged her there for her first day of class. On November 10, 1966, Deborah gave birth to Alfred Jr., who she named after his father, Alfred "Cheetah" Carter, the boy Galen had once been jealous of. Each morning, Bobbette made Deborah's lunch, got her to school, then took care of Alfred all day and most of the night so Deborah could go to class and study. When Deborah graduated, Bobbette made her get her first job—whether Deborah liked it or not, Bobbette was going to help her and that baby.

Deborah's older brothers were doing fine on their own. Lawrence went into business for himself, opening a convenience store in the basement of an old townhouse; Sonny had graduated from high school, joined the air force, and grown into a handsome ladies' man. He did some running around, but pretty much stayed out of trouble. Their younger brother, Joe, was another story.

Authority didn't agree with Joe. He argued with teachers and brawled with other students. He dropped out of school in the seventh grade and ended up in court for "assault by striking" right after his seventeenth birthday. He joined the military at eighteen, but his anger and attitude got him in even more trouble there. He fought his superiors and other soldiers. Sometimes he ended up in the hospital, but more often than not, his fighting landed him in solitary confinement, a dark hole with dirt walls ominously similar to the basement where Ethel once locked him as a child. He preferred being in the hole because it meant no one would bother him. As soon as they let him out, he'd fight another soldier or get belligerent with an officer and they'd throw him back in. He spent nine months in the service, most of it sitting in the hole, growing angrier and angrier. After multiple psychiatric evaluations and treatments, Joe was discharged for an inability to adjust emotionally to military life.

His family had hoped the military would help control his rage and teach him some discipline and respect for authority. Instead, he came out of the military angrier than ever.

A week or so after Joe got home from the military, a tall, skinny neighborhood kid named Ivy walked up to him with a knife and asked if he wanted to start something. Most people wouldn't have done that. At nineteen, Joe was at least four inches shorter than Ivy and only 155 pounds, but people in the neighborhood called him Crazy Joe because he seemed to enjoy violence. Ivy didn't care. He'd been drinking heavily and shooting heroin for years, and he was covered in scars from fighting. He told Joe he was going to kill him.

Joe ignored Ivy the first time. Then, about three months later, on September 12, 1970, Joe was walking down an East Baltimore street with his friend June. It was Saturday night, they'd been drinking, and they'd just started talking up a group of young girls when three other men walked up the street toward them. One of those men was Eldridge Lee Ivy.

When Ivy saw Joe and June talking to the girls, he yelled, saying one of them was his cousin, and they'd better stop messing with her.

"I'm tired of your junk," June yelled back.

The two started arguing, and when Ivy threatened to punch June in the face, Joe jumped between them, calmly telling Ivy he would do no such thing.

Ivy grabbed Joe by the neck, choking him while his two friends tried to pull him off. Joe kicked and yelled, "I'm going to kill your motherfuckin ass!" But Ivy beat him bloody while June watched, terrified.

That night, Joe knocked on Deborah's door. He stared ahead, covered in blood, eyes burning with hate as she cleaned his face and put him on her couch to sober up with some ice packs. He glared at the wall all night, looking scarier and angrier than Deborah had ever seen a person look.

The next morning, Joe went into Deborah's kitchen and took her good carving knife with the black wood handle. Two days later, on September 15, 1970, Joe went to work at his job driving for a local trucking company. By five o'clock, he and a coworker had shared a fifth of Old Granddad whiskey, then another pint. It was still daylight out when Joe got off work and walked to the corner of Lanvale and Montford Avenues in East Baltimore, where Ivy stood on the front stoop of his house, talking to some friends. Joe crossed the street and said, "Hi Ivy," then stabbed him in the chest with Deborah's knife. The blade went straight through Ivy's heart. He staggered down the street and into a neighbor's house with Joe close behind, then collapsed facedown into a pool of his own blood, yelling, "Oh, I'm dying—call an ambulance!" But it was too late. When a fireman arrived a few minutes later, Ivy was dead.

Joe walked away from the murder scene, dropped the knife in a nearby alley, and headed to a pay phone to call his father, but the police had beaten him to it. They'd told Day his son had killed a boy. Sonny and Lawrence told their father to get Joe to Clover, back to the tobacco farms, where he could hide from the law and be safe. Deborah said they were crazy.

"He's got to turn himself in," she told them. "The police got a warrant out saying he wanted dead or alive."

But the men didn't listen. Day gave Joe twenty dollars and put him on a Trailways bus to Clover.

In Lacks Town, Joe drank all day, picked fights with his cousins, and threatened to kill several of them, including Cootie. By the end of Joe's first week, Cootie called Day saying somebody had better come get Joe before he killed someone else or got himself shot. Sonny borrowed Day's car, picked Joe up in Clover, and took him to D.C. to stay with a friend. But Joe couldn't get along there either. The next morning he called Sonny and said, "Come pick me up, I want to turn myself in."

On the morning of September 29, 1970, Joe walked into the Baltimore police headquarters and calmly said, "I'm Joe Lacks. I'm wanted cause I killed Ivy." Then he filled out this form:

Is the defendant employed?	No
Cash on hand or in banks?	Zero
Name of parents?	David Lacks
Have they been to see you?	No
Do you have any friends or members of your family that can get you an attorney?	No. Can't afford one

After that, Joe waited. He knew he was going to plead guilty — he just wanted to get on with it. After five months awaiting trial in a cell, Joe wrote this letter to the criminal court judge:

Dear Sir or Your Honor,
     In the most critical time on this earth is now on this atmosphere today of my missteak no I'll say wronge

comprehendion of corruption that I've place on myself.
A very misslead problem that was not ment to be. Feel
so frustration in making me obnoxious within me,
Asking for a (speedly trial) to Let me know what lays
ahead in the future, I feel as thod I sure be castigate or
chastise for the wronge I've did, So I'm ready to get it
over now with it.

Joe Lacks

(Speedly trial)

(Thank you)

(Your Honor)

Finally, on April 6, 1971—seven months after Ivy's death—Joe
stood in a courtroom and pleaded guilty to murder in the second degree, with Sonny watching nearby. The judge warned Joe repeatedly
that a guilty plea meant waiving his right to a trial, his right to testify,
and his right to appeal her ruling. As the judge spoke, he said "yes
ma'am" and "no ma'am." He told her the alcohol had made him do it
and that he hadn't meant to kill Ivy.

"I tried to hit on top of his shoulder, and he panicked and turned
and caught it in the chest," Joe said. "I was trying to wound him so I
wouldn't let him hurt me. . . . He told me he was going to kill me that
Saturday night me and him got into the argument. I just hope you see
I was trying to protect my life. I was not really wanting any trouble
out of no one at all."

But Ivy's fourteen-year-old neighbor, who'd seen the whole thing,
said Joe had walked right up and stabbed Ivy in the chest, then tried to
stab him again in the back as he staggered away.

When Joe stepped from the stand, his court-appointed lawyer approached the judge to make this final point:

The only thing I would add, Your Honor, is that I talked to his
brother about the young man, and the problem that he also had
in the Army, is a problem that possibly got him into the situa-

tion he is in Court for today. For some reason, somewhere in his life, he has gotten an inferiority complex. And it seems to be a sizable one. It seems that whenever he is confronted by any individual, he sort of takes it rather aggressively, more so than the average individual . . . for the record, [he] had some psychiatric help in the service, but he has never been in any hospital.

Without knowing anything about Joe's life or the abuse he experienced as a child, his lawyer said, "He feels it more necessary to protect himself than the average individual. And possibly, this sets him off, where it would not set off the average person."

"Do people call you Crazy Joe?" the judge asked.

"There was a few friends that called me that," Joe said.

"Do you know why they call you that?"

"No ma'am," he said.

The judge accepted Joe's guilty plea, but asked to see medical and psychiatric reports before deciding his sentence. Those records are sealed, but whatever they contained led her to give him a sentence of only fifteen years out of a possible thirty. The state sent Joe to the Maryland Correctional Institution in Hagerstown, a medium-security prison about seventy-five miles west of Baltimore.

In the beginning, Joe spent his time in prison much as he'd spent it in the military: in the hole for insubordination and fighting. But eventually he stopped fighting and focused his energy inward. Joe found Islam and began spending all his time studying the Koran in his cell. Soon he changed his name to Zakariyya Bari Abdul Rahman.

Meanwhile, on the outside, things were looking pretty good for the other Lacks brothers. Sonny had just been honorably discharged from the Air Force, and Lawrence had a good job working for the railroad. But things weren't so good for Deborah. By the time Zakariyya ended up in prison, Deborah had married Cheetah in a blue chiffon dress in Bobbette and Lawrence's living room. She was eighteen. When Deborah and Cheetah first met, he threw a bowling ball at her on the sidewalk in front of her house. She thought he was playing, but things only got worse after they married. Soon after their second child, LaTonya, was born, Cheetah fell into drugs and started beating Deborah when he was high. Then he started running the streets, disappearing with other women for nights on end, and coming back only to sell drugs out of the house while Deborah's children sat and watched.

One day, as Deborah stood at the sink doing dishes, her hands covered in soap bubbles, Cheetah ran into the kitchen yelling something about her sleeping around on him. Then he smacked her.

"Don't do that again," Deborah said, standing stone-still, her hands still in the dishwater.

Cheetah grabbed a plate from the drying rack and broke it across the side of her face.

"Don't put your hand back on me no more!" Deborah screamed, her hand shooting out of the dishwater, gripping a serrated steak knife.

Cheetah raised his arm to hit her again, but he was clumsy from the drugs and booze. Deborah blocked him with her empty hand and pinned him against the wall. She stuck the tip of the knife into his chest just deep enough to break the skin, then dragged it down past his navel as Cheetah screamed, calling her crazy.

He left her alone for a few days after that, but eventually came home drunk and high and started beating her again. As Cheetah kicked her one night in the living room, Deborah yelled, "Why you always have to be arguing and fussing with me?" When he didn't answer, Deborah decided right then she wanted him dead. He turned and staggered toward the stairs of their apartment, still yelling, and Deborah pushed him hard as she could. He tumbled to the bottom, where he lay bleeding. Deborah stared at him from the top of the stairs, feeling nothing—no fear, no emotion. When he moved, she walked down the steps and dragged him through their basement onto the sidewalk outside. It was the middle of winter and snowing. Deborah dropped him on the ground in front of the house without a coat, slammed the door, and went upstairs to sleep.

The next morning she woke up hoping he'd frozen to death, but instead he was sitting on their front stoop, bruised and cold.

"I feel like some guys jumped me and beat me up," he told her.

She let him in the house and got him washed and fed, all the while thinking what a damn fool he was. While Cheetah slept it off, Deborah called Bobbette, saying, "This is it, he gonna die tonight."

"What are you talking about?" Bobbette asked.

"I got the monkey wrench," Deborah said. "I'm gonna splatter his brains all over the wall. I'm sick of it."

"Don't do it, Dale," Bobbette said. "Look where it got Zakariyya—he's in jail. You kill that man, then what about your children? Now get that monkey wrench outta there."

The next day, after Cheetah left for work, a moving van pulled up to the house. Deborah took the children and everything they owned, then hid at her father's house until she could find her own apartment. As Deborah worked two jobs and struggled to settle into her new life as a single mother, she had no idea she was about to get news that would be harder to handle than anything Cheetah had done.

20

The HeLa
Bomb

In September 1966, a geneticist named Stanley Gartler walked up to the podium at a hotel in Bedford, Pennsylvania. There, in front of George Gey and the other giants of cell culture, Gartler announced that he'd found a "technical problem" in their field.

He was at the Second Decennial Review Conference on Cell Tissue and Organ Culture with more than seven hundred other scientists. They'd come from biotech companies and academia; they'd traveled from New York, England, the Netherlands, Alaska, Japan, and everywhere between to discuss the future of cell culture. The room buzzed with excitement as everyone talked about cell cloning and hybrids, mapping human genes, and using cultures to cure cancer.

Few there had heard of Stanley Gartler, but that was about to change. Gartler leaned into the microphone and told the audience that, in the process of looking for new genetic markers for his research, he'd found that eighteen of the most commonly used cell cultures had one thing in common: they all contained a rare genetic marker called glucose-6-phosphate dehydrogenase-A (G6PD-A), which was present almost exclusively in black Americans. And even among them it was fairly rare.

"I have not been able to ascertain the supposed racial origin of all eighteen lines," Gartler told the audience. "It is known, however, that at least some of these are from Caucasians, and that at least one, HeLa, is from a Negro." He knew this, because a few months earlier, he'd written George Gey:

> I am interested in the racial origin of the person from whom your HeLa cell line was initiated. I have checked a number of the early papers describing the development of the HeLa cell line but have not been able to find any information pertaining to the race of the donor.

When Gey responded that HeLa cells had come from "a colored woman," Gartler knew he'd found the source of the problem.

"It seems to me the simplest explanation," he told the audience, "is that they are all HeLa cell contaminants."

Scientists knew they had to keep their cultures free from bacterial and viral contamination, and they knew it was possible for cells to contaminate one another if they got mixed up in culture. But when it came to HeLa, they had no idea what they were up against. It turned out Henrietta's cells could float through the air on dust particles. They could travel from one culture to the next on unwashed hands or used pipettes; they could ride from lab to lab on researchers' coats and shoes, or through ventilation systems. And they were strong: if just *one* HeLa cell landed in a culture dish, it took over, consuming all the media and filling all the space.

Gartler's findings did not go over well. In the fifteen years since George Gey had first grown HeLa, the number of published articles involving cell culture had more than tripled each year. Scientists had spent millions of dollars conducting research on those cells to study the behavior of each tissue type, comparing one to another, testing the unique responses of different cell types to specific drugs, chemicals, or environments. If all those cells were in fact HeLa, it would mean that millions of dollars had been wasted, and researchers who'd found that various cells behaved differently in culture could have some explaining to do.

Years later, Robert Stevenson, who became president of the American Type Culture Collection, described Gartler's talk to me this way: "He showed up at that meeting with no background or anything else in cell culture and proceeded to drop a turd in the punch bowl."

Stevenson and other members of the Cell Culture Collection Committee sat stunned in the audience as Gartler pointed to a chart on the wall listing the eighteen cell lines that had been contaminated by HeLa, along with the names of the people or places he'd gotten them from. At least six of the contaminated lines came from the ATCC. HeLa had penetrated Fort Knox.

At that point, the ATCC's collection had grown to dozens of different types of cells, all guaranteed to be free from viral and bacterial contamination, and tested to ensure that they hadn't been contaminated with cells from another species. But there was no test to see if one human cell had contaminated another. And, to the naked eye, most cells growing in culture look the same.

Now Gartler was essentially telling the audience that all those years researchers thought they were creating a library of human tissues, they'd probably just been growing and regrowing HeLa. He pointed out that a few years earlier, when scientists started taking protective measures against cross-species contamination — such as working under sterile hoods — it had suddenly become harder to grow new cell lines. And in fact, "very few [new human cell lines] have been reported since." Not only that, he said, but there had been no new examples of "so-called spontaneous transformed human cell cultures" since.

Everyone in the audience knew what that meant. On top of saying they'd possibly wasted more than a decade and millions of research dollars, Gartler was also suggesting that spontaneous transformation — one of the most celebrated prospects for finding a cure for cancer — might not exist. Normal cells didn't spontaneously become cancerous, he said; they were simply taken over by HeLa.

Gartler concluded his talk by saying, "Where the investigator has assumed a specific tissue of origin of the cell line, i.e., liver . . . or bone marrow, the work is open to serious question, and in my opinion would be best discarded."

The room sat silent, dumbfounded, until T. C. Hsu, the chair of Gartler's conference session, spoke. Hsu was the University of Texas geneticist whose earlier work with HeLa and other cells had made it possible to discover the correct number of human chromosomes.

"A few years ago I voiced some suspicion about cell-line contamination," Hsu said. "So I am happy about the paper by Dr. Gartler and am also sure he has made many people unhappy."

He was right, and those people quickly began asking questions.

"How long did you keep them in your laboratory?" one scientist asked, suggesting that Gartler had contaminated the cells himself after they arrived in his lab.

"They were analyzed before being grown in my laboratory," Gartler responded.

"They didn't send them to you frozen?" the scientist asked, knowing that contamination could have occurred while they thawed.

Gartler said that didn't matter—the cells didn't have to be thawed to be tested.

Another scientist wanted to know if the similarity Gartler was seeing between cell lines was just the effect of spontaneous transformation making all cells act the same.

Eventually Robert Stevenson of the Cell Culture Collection Committee spoke up, saying, "It looks like more detective work is needed to see . . . whether we are going to have to start all over again to isolate some new human cell lines."

Hsu stepped in and said, "I would like to give particular priority to those who initiated the cell lines, whom Dr. Gartler has attacked. If there is any defense, we would like to hear it."

Harvard's Robert Chang—whose widely used Chang Liver Cell line was listed as a HeLa contaminant on Gartler's chart—glared from his seat. Chang had used those cells to discover enzymes and genes specific to liver cells. If Gartler was right and the cells were actually from Henrietta's cervix, Chang's liver research using them was worthless.

Leonard Hayflick had an especially personal connection with his cell line, WISH, which Gartler had listed as contaminated: he'd grown it using cells from the amniotic sac in which his unborn daughter had once floated. He asked Gartler whether it was possible to find G6PD-A in samples from white people.

"Caucasian subjects with G6PD-A have not been reported," Gartler told him.

Later that day—in a talk chaired by George Gey—Hayflick delivered a paper, on the "facts and theories" of spontaneous transformation of cells in culture. Before beginning his talk, Hayflick stood at the podium and announced that, since WISH cells supposedly tested positive for a genetic marker found only in black people, he'd called his wife during the break to ask if he was, in fact, his daughter's father. "She assured me that my worst fears were unfounded," Hayflick said. The room erupted in laughter, and no one said anything else publicly about Gartler's findings.

But a few people took Gartler seriously: before leaving the conference, Stevenson met several of the top cell culturists for lunch. He told them to go back to their labs after the conference and start testing cells for the G6PD-A genetic marker, to see how widespread this problem might be. Many of their cell lines tested positive, including the skin cells George Hyatt had transplanted onto a soldier's arm years earlier. Since Hyatt had no HeLa cells in his lab at the time, the cells in his experiment must have been contaminated before they arrived. And though few realized it, the same thing was happening in laboratories around the world.

Still, many scientists refused to believe HeLa contamination was real. After the conference where Gartler dropped what became known as "the HeLa bomb," most researchers kept right on working with the cells he'd said were contaminated. But Stevenson and a few other scientists realized the potential scope of the HeLa contamination problem, so they began working to develop genetic tests that could specifically identify HeLa cells in culture instead of just testing for the presence of G6PD-A. And those genetic tests would eventually lead them to Henrietta's family.

21

Night
Doctors

Two months after Sonny Lacks stood me up, I sat waiting for him again, this time in the lobby of the Baltimore Holiday Inn. It was New Year's Day, and he was nearly two hours late. I figured he'd backed out again, so I started packing to leave. Then I heard a man's voice yell, "So you're Miss Rebecca!"

Suddenly, Sonny was standing beside me with a sweet and bashful gap-toothed grin that made him look like a fifty-year-old teenager. He laughed and patted me on the back.

"You just won't give up, will you?" he said. "I got to tell you, only person I know more hardheaded than you is my sister Dale." He grinned and straightened his black driving cap. "I tried to convince her to come meet you today, but she won't listen."

Sonny had a loud laugh and mischievous eyes that squinted nearly closed when he smiled. His face was warm and handsome, open to the world. He was thin, five foot nine at most, with a carefully manicured mustache. He reached for my bag.

"Okay then," he said, "we best get this thing goin."

I followed him to a Volvo he'd left unlocked and idling in the parking lot next to the hotel. He'd borrowed it from one of his daughters. "Nobody wants to ride in my old raggedy van," he said, easing the car into gear. "You ready to go see the Big Kahuna?"

"The Big Kahuna?"

"Yep," he said, grinning. "Deborah says you got to talk to our brother Lawrence before anybody else talk to you. He'll check you out, decide what's what. If he say it's okay, maybe then the rest of us will talk to you."

We drove in silence for several blocks.

"Lawrence is the only one of us kids who remembers our mother," Sonny said eventually. "Deborah and I don't know nothing about her." Then, without looking from the road, Sonny told me everything he knew about his mother.

"Everybody say she was real nice and cooked good," he said. "Pretty too. Her cells have been blowed up in nuclear bombs. From her cells came all these different creations—medical miracles like polio vaccines, some cure for cancer and other things, even AIDS. She liked takin care of people, so it make sense what she did with them cells. I mean, people always say she was really just hospitality, you know, fixing everything up nice, make a good place, get up, cook breakfast for everybody, even if it's twenty of them."

He pulled into an empty alley behind a row of red brick townhouses and looked at me for the first time since we'd gotten in the car.

"This is where we take scientists and reporters wanting to know about our mother. It's where the family gangs up on them," he said, laughing. "But you seem nice, so I'll do you a favor and not go get my brother Zakariyya this time."

I got out of the car and Sonny drove away, yelling, "Good luck!" out the window.

All I knew about Sonny's brothers was that they were angry and one of them had murdered someone—I wasn't sure which one, or why. A few months earlier, when Deborah gave me Lawrence's phone number and swore she'd never talk to me, she'd said, "Brother gets mad when white folks come askin about our mother."

As I walked through a narrow, half-cement yard from the alley to Lawrence's house, a wisp of smoke seeped through the screen door of his kitchen, where static blared from a small television on a folding table. I knocked, then waited. Nothing. I stuck my head into the kitchen, where fat pork chops sat burning on the stove. I yelled hello. Still nothing.

I took a deep breath and walked inside. As I closed the door behind me, Lawrence appeared, seeming bigger than two of me, his 275-pound, six-foot frame spanning the width of the narrow kitchen, one hand on the counter, the other on the opposite wall.

"Well hello there, Miss Rebecca," he said, giving me a once-over. "You wanna taste the meat I cooked?"

It had been a decade or so since I'd eaten pork, but suddenly that seemed irrelevant. "How could I resist?" I said.

A sweet grin spread across Lawrence's face. He was sixty-four, but aside from his gray curls, he seemed decades younger, with smooth hazelnut-brown skin and youthful green eyes. He hiked up his baggy blue jeans, wiped his hands on his grease-stained T-shirt, and clapped.

"Okay then," he said, "that's good. That's real good. I'm gonna fry you up some eggs too. You're too damn skinny."

While he cooked, Lawrence talked about life down in the country. "When older folks went to town to sell tobacco, they'd come back with a piece of bologna for us kids to share. And sometimes if we were good, they'd let us sop up the bacon grease with a piece of bread." His memory for detail was impressive. He drew pictures of the horse-drawn wagon Day had made out of two-by-fours. He showed me, with string and napkins, how he tied tobacco into bundles for drying when he was a child.

But when I asked about his mother, Lawrence fell silent. Eventually he said, "She was pretty." Then he went back to talking about tobacco. I asked about Henrietta again and he said, "My father and his friends used to race horses up and down Lacks Town road." We went in circles like this until he sighed and told me he didn't remember his mother. In fact, he said, he didn't remember most of his teen years.

"I blacked it out of my mind because of the sadness and hurting," he told me. And he had no intention of unblocking it.

"The only memory I have about my mother is her being strict," he told me. He remembered her making him hand-wash diapers in the sink; he'd hang them to dry, then she'd dump them back in the water, saying they weren't clean enough. But the only times she whipped him were for swimming off the pier in Turner Station. "She'd make me go fetch a switch to get a beatin with, then send me back out sayin get a bigger one, then a bigger one, then she'd wrap all them together and haul off on my tail."

As he talked, the kitchen filled with smoke again—we'd both forgotten he was cooking. Lawrence shooed me from the kitchen table into the living room, where he sat me in front of a plastic Christmas place mat with a plate of fried eggs and a chunk of charred pork the size of my hand, only thicker. Then he collapsed into a wooden chair beside me, put his elbows on his knees, and stared at the floor in silence while I ate.

"You're writing a book about my mama," he said finally.

I nodded as I chewed.

"Her cells growin big as the world, cover round the whole earth," he said, his eyes tearing as he waved his arms in the air, making a planet around him. "That's kinda weird . . . They just steady growin and growin, steady fightin off whatever they fightin off."

He leaned forward in his chair, his face inches from mine, and whispered, "You know what I heard? I heard by the year 2050, babies will be injected with serum made from my mama's cells so they can live to eight hundred years old." He gave me a smile like, *I bet your mama can't top that.* "They're going to get rid of disease," he said. "They're a miracle."

Lawrence fell back in his chair and stared into his lap, his smile collapsing. After a long quiet moment, he turned and looked into my eyes.

"Can you tell me what my mama's cells really did?" he whispered. "I know they did something important, but nobody tells us nothing,"

When I asked if he knew what a cell was, he stared at his feet as if I'd called on him in class and he hadn't done his homework.

"Kinda," he said. "Not really."

I tore a piece of paper from my notebook, drew a big circle with a small black dot inside, and explained what a cell was, then told him some of the things HeLa had done for science, and how far cell culture had come since.

"Scientists can even grow corneas now," I told him, reaching into my bag for an article I'd clipped from a newspaper. I handed it to him and told him that, using culturing techniques HeLa helped develop, scientists could now take a sample of someone's cornea, grow it in culture, then transplant it into someone else's eye to help treat blindness.

"Imagine that," Lawrence said, shaking his head. "It's a miracle!"

Suddenly, Sonny threw open the screen door, yelling, "Miss Rebecca still alive in here?" He leaned in the doorway between the kitchen and living room.

"Looks like you passed the test," he said, pointing at my half-empty plate.

"Miss Rebecca telling me about our mother cells," Lawrence said. "She told me fascinating stuff. Did you know our mother cells gonna be used to make Stevie Wonder see?"

"Oh, well, actually, it's not *her* cells being put into people's eyes," I said, stammering. "Scientists are using technology her cells helped develop to grow *other* people's corneas."

"That's a miracle," Sonny said. "I didn't know about that, but the other day President Clinton said the polio vaccine is one of the most important things that happened in the twentieth century, and her cells involved with that too."

"That's a miracle," Lawrence said.

"So is this," Sonny said, slowly spreading his arms and stepping

aside to reveal his eighty-four-year-old father, Day, teetering on un-
steady legs behind him.

Day hadn't left the house in nearly a week because of a nosebleed
that wouldn't stop. Now he stood in the doorway in faded jeans, a
flannel shirt, and blue plastic flip-flops, even though it was January.
He was thin and frail, barely able to hold himself upright. His light
brown face had grown tough with age, cracked but soft, like a pair of
well-worn work boots. His silver hair was covered with a black driving cap identical to Sonny's.

"He's got the gangrene in his feet," Sonny said, pointing to Day's
toes, which were several shades darker than the rest of him and covered with open sores. "His feet hurt too much in regular shoes." Gangrene was spreading from Day's toes to his knee; his doctor said his
toes needed amputating, but Day refused. He said he didn't want doctors cutting on him like they did Henrietta. At fifty-two, Sonny felt
the same way; his doctors said he needed angioplasty, but he swore
he'd never do it.

Day sat beside me, brown plastic sunglasses shading his constantly tearing eyes.

"Daddy," Lawrence yelled, "did you know mama's cells gonna
make Stevie Wonder see?"

Day shook his head in what looked like slow motion. "Nope," he
mumbled. "Didn't know that till just now. Don't surprise me none
though."

Then there was a thump on the ceiling and the rustling of someone walking around, and Lawrence jumped from the table and ran
into the kitchen. "My wife is a fire dragon without morning coffee,"
he said. "I better make some." It was two in the afternoon.

A few minutes later, Bobbette Lacks walked down the stairs and
through the living room slowly, wearing a faded blue terry-cloth robe.
Everyone stopped talking as she passed and headed into the kitchen
without saying a word or looking at anyone.

Bobbette seemed like a loud person being quiet, like a woman with an enormous laugh and temper who might erupt with either at any moment. She exuded *Don't mess with me,* her face stern and staring straight ahead. She knew why I was there, and had plenty to say on the subject, but seemed utterly exhausted at the idea of talking to me, yet another white person wanting something from the family.

She disappeared into the kitchen and Sonny slid a crumpled piece of paper into Day's hand, a printout of the picture of Henrietta with her hands on her hips. He grabbed my tape recorder from the center of the table, handed it to Day, and said, "Okay, Miss Rebecca got questions for you, Pop. Tell her what you know."

Day took the recorder from Sonny's hand and said nothing.

"She just want to know everything Dale always askin you about," Sonny said.

I asked Sonny if maybe he could call Deborah to see if she'd come over, and the Lacks men shook their heads, laughing.

"Dale don't want to talk to nobody right now," Sonny said.

"That's cause she's tired of it," Day grumbled. "They always askin questions and things, she keep givin out information and not gettin nuthin. They don't even give her a postcard."

"Yep," Sonny said, "that's right. All they wants to do is know everything. And that's what Miss Rebecca wants too. So go on Daddy, tell her, just get this over with."

But Day didn't want to talk about Henrietta's life.

"First I heard about it was, she had that cancer," he said, repeating the story he'd told dozens of reporters over the years, almost verbatim. "Hopkins called me, said come up there cause she died. They asked me to let them have Henrietta and I told them no. I said, 'I don't know what you did, but you killed her. Don't keep cuttin on her.' But after a time my cousin said it wouldn't hurt none, so I said okay."

Day clenched his three remaining teeth. "I didn't sign no papers," he said. "I just told them they could do a topsy. Nothin else. Them doctors never said nuthin about keepin her alive in no tubes or growin no cells. All they told me was they wanted to do a topsy see if they could help my children. And I've always just knowed this much: they is the doctor, and you got to go by what they say. I don't know as much as they do. And them doctors said if I gave em my old lady, they could use her to study that cancer and maybe help my children, my grandchildren."

"Yeah!" Sonny yelled. "They said it would help his kids in case they come down with cancer. He had five kids, what was he going to do?"

"They knew them cells was already growin when I come down there after she died," Day said, shaking his head. "But they didn't tell me nuthin bout that. They just asked if they could cut her up see about that cancer."

"Well what do you expect from Hopkins?" Bobbette yelled from the kitchen, where she sat watching a soap opera. "I wouldn't even go there to get my toenails cut."

"Mmm hmm," Day yelled back, thumping his silver cane on the floor like an exclamation point.

"Back then they did things," Sonny said. "Especially to black folks. John Hopkins was known for experimentin on black folks. They'd snatch em off the street . . ."

"That's right!" Bobbette said, appearing in the kitchen door with her coffee. "Everybody knows that."

"They just snatch em off the street," Sonny said.

"Snatchin people!" Bobbette yelled, her voice growing louder.

"Experimentin on them!" Sonny yelled.

"You'd be surprised how many people disappeared in East Baltimore when I was a girl," Bobbette said, shaking her head. "I'm telling you, I lived here in the fifties when they got Henrietta, and we weren't allowed to go anywhere near Hopkins. When it got dark and we were young, we had to be *on the steps,* or Hopkins might get us."

The Lackses aren't the only ones who heard from a young age that Hopkins and other hospitals abducted black people. Since at least the 1800s, black oral history has been filled with tales of "night doctors" who kidnapped black people for research. And there were disturbing truths behind those stories.

Some of the stories were conjured by white plantation owners taking advantage of the long-held African belief that ghosts caused disease and death. To discourage slaves from meeting or escaping, slave owners told tales of gruesome research done on black bodies, then covered themselves in white sheets and crept around at night, posing as spirits coming to infect black people with disease or steal them for research. Those sheets eventually gave rise to the white hooded cloaks of the Ku Klux Klan.

But night doctors weren't just fictions conjured as scare tactics. Many doctors tested drugs on slaves and operated on them to develop new surgical techniques, often without using anesthesia. Fear of night doctors only increased in the early 1900s, as black people migrated north to Washington, D.C., and Baltimore, and news spread that medical schools there were offering money in exchange for bodies. Black corpses were routinely exhumed from graves for research, and an underground shipping industry kept schools in the North supplied with black bodies from the South for anatomy courses. The bodies sometimes arrived, a dozen or so at a time, in barrels labeled *turpentine.*

Because of this history, black residents near Hopkins have long believed the hospital was built in a poor black neighborhood for the benefit of scientists—to give them easy access to potential research subjects. In fact, it was built for the benefit of Baltimore's poor.

Johns Hopkins was born on a tobacco plantation in Maryland where his father later freed his slaves nearly sixty years before Emancipation. Hopkins made millions working as a banker and grocer, and selling his own brand of whiskey, but he never married and had no children. So in 1873, not long before his death, he donated $7 million to start a medical school and charity hospital. He wrote a letter to the twelve men he'd chosen to serve as its board of trustees, outlining his wishes. In it he explained that the purpose of Hopkins Hospital was to help those who otherwise couldn't get medical care:

The indigent sick of this city and its environs, without regard to sex, age, or color, who require surgical or medical treatment, and who can be received into the hospital without peril to other inmates, and the poor of the city and State, of all races, who are stricken down by any casualty, shall be received into the hospital without charge.

He specified that the only patients to be charged were those who could easily afford it, and that any money they brought in should then be spent treating those without money. He also set aside an additional $2 million worth of property, and $20,000 in cash each year, specifically for helping black children:

> It will be your duty hereafter to provide . . . suitable buildings for the reception, maintenance and education of orphaned colored children. I direct you to provide accommodations for three or four hundred children of this class; you are also authorized to receive into this asylum, at your discretion, as belonging to such class, colored children who have lost one parent only, and in exceptional cases to receive colored children who are not orphans, but may be in such circumstances as to require the aid of charity.

Hopkins died not long after writing that letter. His board of trustees—many of them friends and family—created one of the top medical schools in the country, and a hospital whose public wards provided millions of dollars in free care to the poor, many of them black.

But the history of Hopkins Hospital certainly isn't pristine when it comes to black patients. In 1969, a Hopkins researcher used blood samples from more than 7,000 neighborhood children—most of them from poor black families—to look for a genetic predisposition to criminal behavior. The researcher didn't get consent. The American Civil Liberties Union filed suit claiming the study violated the boys' civil rights and breached confidentiality of doctor-patient relationships by releasing results to state and juvenile courts. The study was halted, then resumed a few months later using consent forms.

And in the late nineties, two women sued Hopkins, claiming that its researchers had knowingly exposed their children to lead, and hadn't promptly informed them when blood tests revealed that their children had elevated lead levels—even when one developed lead poisoning. The research was part of a study examining lead abatement methods, and all families involved were black. The researchers had treated several homes to varying degrees, then encouraged landlords to rent those homes to families with children so they could then monitor the children's lead levels. Initially, the case was dismissed. On appeal, one judge compared the study to Southam's HeLa injections, the Tuskegee study, and Nazi research, and the case eventually settled out of court. The Department of Health and Human Services launched an investigation and concluded that the study's consent forms "failed to provide an adequate description" of the different levels of lead abatement in the homes.

But today when people talk about the history of Hopkins's relationship with the black community, the story many of them hold up as the worst offense is that of Henrietta Lacks—a black woman whose body, they say, was exploited by white scientists.

Sitting in Lawrence's living room, Sonny and Bobbette yelled back and forth for nearly an hour about Hopkins snatching black people. Eventually, Sonny leaned back in his chair and said, "John Hopkin didn't give us no information about anything. That was the bad part. Not the sad part, but the bad part, cause I don't know if they didn't give us information because they was making money out of it, or if they was just wanting to keep us in the dark about it. I think they made money out of it, cause they were selling her cells all over the world and shipping them for dollars."

"Hopkins say they gave them cells away," Lawrence yelled, "but they made millions! It's not fair! She's the most important person in the world and her family living in poverty. If our mother so important to science, why can't we get health insurance?"

Day had prostate cancer and asbestos-filled lungs. Sonny had a bad heart, and Deborah had arthritis, osteoporosis, nerve deafness, anxiety, and depression. With all that plus the whole family's high blood pressure and diabetes, the Lackses figured they pretty much supported the pharmaceutical industry, plus several doctors. But their insurance came and went. Some were covered through Medicare, others on and off by spouses, but they all went stretches with no coverage or money for treatment.

As the Lacks men talked about Hopkins and insurance, Bobbette snorted in disgust and walked to her recliner in the living room. "My pressure's goin up and I'm not gonna die over this, you know?" The whole thing just wasn't worth getting riled up over, she said. But she couldn't help herself. "Everybody knew black people were disappearing cause Hopkins was experimenting on them!" she yelled. "I believe a lot of it was true."

"Probably so," Sonny said. "A lot might a been myth too. You never know. But one thing we do know, them cells about my mother ain't no myth."

Day thumped his cane again.

"You know what *is* a myth?" Bobbette snapped from the recliner. "Everybody always saying Henrietta Lacks donated those cells. She didn't donate nothing. They took them and didn't ask." She inhaled a deep breath to calm herself. "What really would upset Henrietta is the fact that Dr. Gey never told the family anything—we didn't know nothing about those cells and he didn't care. That just rubbed us the wrong way. I just kept asking everybody, 'Why didn't they say anything to the family?' They knew how to contact us! If Dr. Gey wasn't dead, I think I would have killed him myself."

22

"The Fame She So Richly Deserves"

One afternoon in the late spring of 1970, George Gey stood in his favorite waders on the bank of the Potomac River, where he and several other Hopkins researchers had been fishing together every Wednesday for years. Suddenly Gey was so exhausted, he could hardly hold his fishing rod. His buddies dragged him up the embankment to the white Jeep he'd bought using money from a cancer research award.

Soon after that fishing trip, at the age of seventy-one, Gey learned he had the disease he'd spent his entire life trying to fight. And he had one of its most deadly forms: pancreatic cancer. If doctors didn't operate, Gey knew he would die within months. If they did, it might buy him a little time. Or it might not.

On August 8, 1970, around 6:00 a.m., Margaret called each member of the Gey lab's staff, including a postdoctoral student who'd just flown in on a red-eye from Europe.

"Come down to the lab as fast as you can," she told them. "There's going to be an emergency procedure this morning." She didn't tell them what that procedure would be.

Before going into the operating room, George told his surgeons that he wanted them to take samples of his tumor, just as Dr. Wharton had done with Henrietta's tumor decades earlier. Gey gave his lab staff careful instructions for growing GeGe, a line of cancer cells taken from his pancreas. He hoped that his cells, like Henrietta's, would become immortal.

"Work all day and night if you have to," he told his postdocs and assistants. "Make this happen."

Soon, with Gey anesthetized on the operating table, surgeons opened him up and found that the cancer was inoperable—growths covered his stomach, spleen, liver, and intestines. They worried that cutting into the cancer might kill him. Despite Gey's wishes, they sewed him up without taking any samples. When he awoke from anesthesia and found out there would be no GeGe line, he was furious. If this cancer was going to kill him, he wanted it to help advance science in the process.

As soon as he'd recovered enough from his surgery to travel, Gey began contacting cancer researchers around the country, asking who was doing research on pancreatic cancer and needed a patient to experiment on. He was flooded with replies—some from scientists he didn't know, others from friends and colleagues.

In the three months between his surgery and his death, Gey went to the Mayo Clinic in Minnesota for a week of treatments with an experimental Japanese drug that made him violently ill. His son, George Jr., who had just finished medical school, sat with Gey through the whole thing and made sure he had a freshly pressed suit each day. After leaving the Mayo Clinic, Gey spent several days in New York City at Sloan-Kettering for another study, and he underwent chemotherapy at Hopkins using a drug not yet approved for use in humans.

Gey was six and a half feet tall and about 215 pounds when he was diagnosed, but he withered quickly. He often doubled over from abdominal pain, he vomited constantly, and the treatments soon left him confined to a wheelchair. But he continued showing up at the lab and writing letters to his colleagues. At some point not long before his

death, he told his former assistant Mary Kubicek that it was fine to release Henrietta's name if anyone asked, since it had been so many
years. But Mary never told a soul.

George Gey died on November 8, 1970.

A few months after Gey's death, Howard Jones
and several Hopkins colleagues—including Victor McKusick, a leading
geneticist—decided to write an article about the history of the HeLa
cell line as a tribute to Gey's career. Before writing the article, Jones
pulled Henrietta's medical records to remind himself of the details of
her case. When he saw the photographs of her biopsy, he immediately
realized her tumor had been misdiagnosed. To be sure, he dug out the
original biopsy sample, which had been stored on a shelf since 1951.

In December 1971, when Jones and his colleagues published their
tribute to Gey in the journal *Obstetrics and Gynecology,* they reported
that the original pathologist had "misinterpreted" and "mislabeled"
Henrietta's cancer. Her tumor was invasive, but not an epidermoid carcinoma as originally diagnosed. Rather, the article said, it was "a very
aggressive adenocarcinoma of the cervix," meaning it originated from
glandular tissue in her cervix instead of epithelial tissue.

A misdiagnosis of this type was not uncommon at the time. In
1951, the same year Jones biopsied Henrietta's tumor, researchers
from Columbia University reported that the two types of cancer were
easily and often confused.

According to Howard Jones and other gynecologic oncologists
I talked with, the correct diagnosis wouldn't have changed the way
Henrietta's cancer was treated. By 1951, at least twelve studies had
found that cervical adenocarcinomas and epidermoid carcinomas responded the same to radiation, which was the treatment of choice for
both types.

Though it wouldn't have changed Henrietta's treatment, this new
diagnosis could help explain why the cancer spread throughout her
body so much faster than her doctors expected. Cervical adenocarci-

nomas are often more aggressive than epidermoid. (Her syphilis, it turns out, could have been a factor as well—syphilis can suppress the immune system and allow cancer to spread faster than normal.)

Regardless, Jones and his colleagues wrote, the new diagnosis was "but a footnote to the abiding genius of George Gey. . . . It has been often said that scientific discovery results when the right man is in the right place at the right time." Gey, they said, was precisely that man. And HeLa was the result of that luck. "If allowed to grow uninhibited under optimal cultural conditions, [HeLa] would have taken over the world by this time," they wrote. "The biopsy . . . has secured for the patient, Henrietta Lacks as HeLa, an immortality which has now reached 20 years. Will she live forever if nurtured by the hands of future workers? Even now Henrietta Lacks, first as Henrietta and then as HeLa, has a combined age of 51 years."

This was the first time Henrietta's real name appeared in print. Along with it, for the first time, ran the now ubiquitous photograph of Henrietta standing with her hands on her hips. The caption called her "Henrietta Lacks (HeLa)." With that publication, Henrietta's doctor and his colleagues forever linked Henrietta, Lawrence, Sonny, Deborah, Zakariyya, their children, and all future generations of Lackses to the HeLa cells, and the DNA inside them. And Henrietta's identity would soon spread from lab to lab as quickly as her cells.

Just three weeks after Henrietta's name was first published, Richard Nixon signed the National Cancer Act into law and launched the War on Cancer, designating $1.5 billion for cancer research over the next three years. In a move many believe was intended to distract attention from the Vietnam War, Nixon announced that scientists would cure cancer within five years, just in time for the United States Bicentennial.

With this new funding came intense political pressure for scientists to meet the president's deadline. Researchers raced to find what they believed to be the elusive cancer virus, with hopes of developing a vaccine to prevent it. And in May 1972, Nixon pledged that American and Russian scientists would work together in a biomedical exchange program to find the virus.

Though much of the War on Cancer hinged on research using cell cultures, few people knew that those cultures had been contaminated with HeLa. A *Washington Post* reporter had been at the conference when Gartler announced the contamination problem, but he hadn't covered it, and most scientists were still denying that the problem existed. Some were even conducting studies aimed at disproving Gartler's findings.

But the problem wasn't going to go away. Near the end of 1972, when Russian scientists claimed they'd found a cancer virus in cells from Russian cancer patients, the U.S. government had samples of the cells hand-delivered to the Naval Biomedical Research Laboratory in California for testing. It turned out those cells weren't from Russian cancer patients at all. They were from Henrietta Lacks.

The man who discovered that fact was Walter Nelson-Rees, a chromosome expert who was director of cell culture at the Naval laboratory. Nelson-Rees had been in the audience when Gartler presented his infamous research, and he was one of the few scientists who believed it. Nelson-Rees had since been hired by the National Cancer Institute to help stop the contamination problem. He would become known as a vigilante who published "HeLa Hit Lists" in *Science,* listing any contaminated lines he found, along with the names of researchers who'd given him the cells. He didn't warn researchers when he found that their cells had been contaminated with HeLa; he just published their names, the equivalent of having a scarlet *H* pasted on your lab door.

Despite all the evidence, most researchers still refused to believe there was a problem. And the media didn't seem to notice, until news hit that the Russian cells had been contaminated by American ones. Only then did newspapers in London, Arizona, New York, and Washington run headlines saying things like CANCER CELLS FROM LONG-DEAD WOMAN INVADE OTHER CULTURES. They reported "serious confusion," "misguided research," and millions of wasted dollars.

Suddenly, for the first time since the *Collier's* article in the fifties, the press was very interested in the woman behind those cells. They wrote about her "unusual kind of immortality" in one article after another; they called her Helen Larsen or Helen Lane, but never Henrietta Lacks, because Jones and McKusick had published her name in a small science journal few people read.

Rumors spread about the identity of this mysterious Helen L. Some said she'd been Gey's secretary, or maybe his mistress. Others said she was a prostitute off the streets near Hopkins or a figment of Gey's imagination, a fictitious character he'd created to hide the true identity of the woman behind the cells.

As Helen showed up in articles again and again with different last names, a few scientists began feeling the need to set the record straight. On March 9, 1973, the journal *Nature* published a letter from J. Douglas, a biologist at Brunel University:

> It is twenty-one years since George Gey established the famous HeLa cells in culture. It has been estimated that the weight of these cells in the world today exceeds that of the American negro from whose cervical tumour they originated. That lady has achieved true immortality, both in the test-tube and in the hearts and minds of scientists the world over, since the value of HeLa cells in research, diagnosis, etc., is inestimable. Yet we do not know her name! It has been widely stated that He and La are the first letters of her names but whereas one textbook says the names were Helen Lane another says Henrietta Lacks. My letters to the authors, inquiring the source of their information, like the letter to the hospital from which Gey's paper emanated, remain unanswered. Does anyone know for sure? Would it be contrary to medical ethics in the HeLa cell's coming-of-age year to authenticate the name and let He . . . La . . . enjoy the fame she so richly deserves?

Douglas was flooded with responses. There's no record of readers addressing his question about medical ethics, but they did correct his grammar and his use of the word "negro" in place of "negress." Many replies offered the names of women they believed were behind the HeLa cells: Helga Larsen, Heather Langtree, even the actress Hedy Lamarr. In a follow-up letter on April 20, 1973, Douglas announced that all those women should "withdraw as gracefully as they can," because he'd received a letter from Howard W. Jones that left "no doubt that HeLa cells were named after Henrietta Lacks."

And Jones wasn't the only one setting the record straight about Henrietta's name: soon Victor McKusick, one of Jones's coauthors, would send a similar letter to a reporter from *Science,* correcting her misuse of the name Helen Lane. In response, the journalist wrote a short follow-up article in *Science* titled "HeLa (for Henrietta Lacks)." In it she explained that she'd inadvertently "repeated the lore about the origin of those cells." Then, in one of the most widely read science journals in the world, she corrected her error: "Helen Lane, it seems, never lived. But Henrietta Lacks did, long protected by the pseudonym Helen Lane." She also reported that Henrietta's tumor had been incorrectly diagnosed.

"None of this alters the validity of the work done with HeLa cells," she wrote, "but it may be worth noting—for the record."

Part Three

IMMORTALITY

23

"It's Alive"

On a hazy day in 1973, in a brown brick row house five doors down from her own, Bobbette Lacks sat at her friend Gardenia's dining room table. Gardenia's brother-in-law was in town from Washington, D.C., and they'd all just finished having lunch. As Gardenia clanked dishes in the kitchen, her brother-in-law asked Bobbette what she did for a living. When she told him she was a patient aide at Baltimore City Hospital, he said, "Really? I work at the National Cancer Institute."

They talked about medicine and Gardenia's plants, which covered the windows and counters. "Those things would die in my house," Bobbette said, and they laughed.

"Where you from anyway?" he asked.

"North Baltimore."

"No kidding, me too. What's your last name?"

"Well, it was Cooper, but my married name is Lacks."

"Your last name is Lacks?"

"Yeah, why?"

"It's funny," he said, "I've been working with these cells in my lab for years, and I just read this article that said they came from a woman named Henrietta Lacks. I've never heard that name anywhere else."

Bobbette laughed. "My mother-in-law's Henrietta Lacks but I know you're not talking about her—she's been dead almost twenty-five years."

"Henrietta Lacks is your mother-in-law?" he asked, suddenly excited. "Did she die of cervical cancer?"

Bobbette stopped smiling and snapped, "How'd you know that?"

"Those cells in my lab have to be hers," he said. "They're from a black woman named Henrietta Lacks who died of cervical cancer at Hopkins in the fifties."

"What?!" Bobbette yelled, jumping up from her chair. "What you mean you got her cells in your lab?"

He held his hands up, like *Whoa, wait a minute.* "I ordered them from a supplier just like everybody else."

"What do you mean, 'everybody else'?!" Bobbette snapped. "*What* supplier? Who's got cells from my mother-in-law?"

It was like a nightmare. She'd read in the paper about the syphilis study at Tuskegee, which had just been stopped by the government after forty years, and now here was Gardenia's brother-in-law, saying Hopkins had part of Henrietta alive and scientists everywhere were doing research on her and the family had no idea. It was like all those terrifying stories she'd heard about Hopkins her whole life were suddenly true, and happening to her. *If they're doing research on Henrietta,* she thought, *it's only a matter of time before they come for Henrietta's children, and maybe her grandchildren.*

Gardenia's brother-in-law told Bobbette that Henrietta's cells had been all over the news lately because they'd been causing problems by contaminating other cultures. But Bobbette just kept shaking her head and saying, "How come nobody told her family part of her was still alive?"

"I wish I knew," he said. Like most researchers, he'd never thought about whether the woman behind HeLa cells had given them voluntarily.

Bobbette excused herself and ran home, bursting through the screen door into the kitchen, yelling for Lawrence, "Part of your mother, it's alive!"

Lawrence called his father to tell him what Bobbette had heard, and Day didn't know what to think. *Henrietta's alive?* he thought. It didn't make any sense. He'd seen her body at the funeral in Clover himself. Did they go dig it up? Or maybe they did something to her during that autopsy?

Lawrence called the main switchboard at Hopkins, saying, "I'm calling about my mother, Henrietta Lacks—you got some of her alive in there." When the operator couldn't find a record of a patient named Henrietta Lacks in the hospital, Lawrence hung up and didn't know who else to call.

Soon after Lawrence called Hopkins, in June 1973, a group of researchers gathered around a table at Yale University at the First International Workshop on Human Gene Mapping, a first step toward the Human Genome Project. They were talking about how to stop the HeLa contamination problem, when someone pointed out that the whole mess could be sorted out if they found genetic markers specific to Henrietta and used them to identify which cells were hers and which weren't. But doing that would require DNA samples from her immediate family—preferably her husband as well as her children—to compare their DNA to HeLa's and create a map of Henrietta's genes.

Victor McKusick, one of the scientists who'd first published Henrietta's name, happened to be at that table. He told them he could help. Henrietta's husband and children were still patients at Hopkins, he said, so finding them wouldn't be difficult. As a physician on staff, McKusick had access to their medical records and contact information.

The geneticists at the conference were thrilled. If they had access to DNA from Henrietta's children, they could not only solve the contamination problem but also study Henrietta's cells in entirely new ways. McKusick agreed, so he turned to one of his postdoctoral fellows, Susan Hsu, and said, "As soon as you get back to Baltimore, get this done."

McKusick didn't give Hsu instructions for explaining the research to the Lackses. All she knew was that Victor McKusick had told her to call the family.

"He was like a god," Hsu told me years later. "He was a famous, famous man, he trained most of the other famous medical geneticists in the world. When Dr. McKusick said, 'You go back to Baltimore, get this blood drawn,' I did it."

When Hsu got home from the conference, she called Day to ask if she could draw blood from his family. "They said they got my wife and she part alive," he told me years later. "They said they been doin experiments on her and they wanted to come test my children see if they got that cancer killed their mother."

But Hsu hadn't said anything about testing the children for cancer. There was no such thing as a "cancer test," and even if there had been, McKusick's lab wouldn't have been doing one, because he wasn't a cancer researcher. McKusick was a renowned geneticist who'd founded the world's first human genetics department at Hopkins, where he maintained a catalog of hundreds of genes, including several he'd discovered himself in Amish populations. He compiled information about known genes and the research done on them into a database called *Mendelian Inheritance in Man,* the bible of the field, which now has nearly twenty thousand entries and is still growing.

McKusick and Hsu were hoping to use somatic-cell hybridization to test the Lacks family for several different genetic markers, including specific proteins called *HLA markers.* By testing Henrietta's children, they hoped to find out what Henrietta's HLA markers might have been, so they could use those to identify her cells.

Hsu had come to America from China, and English wasn't her native language. According to Hsu, when she called Day in 1973, she told him this: "We come to draw blood to get HLA antigen, we do genetic marker profile because we can deduce a lot of Henrietta Lacks genotype from the children and the husband."

When I asked her if Day seemed to understand, Hsu said, "They are very receptible to us when I made phone call. They are pretty intelligent. I think Mr. Lacks pretty much already knew that his wife made a contribution and are very aware of the value of HeLa cells. They probably heard people talking that the cell line is such important thing. Everybody talking about HeLa back then. They are a very nice family, so they very nicely let us draw blood."

Hsu's accent was strong, and so was Day's—he spoke with a Southern country drawl so thick his own children often had a hard time understanding him. But language wasn't their only barrier. Day wouldn't have understood the concept of immortal cells or HLA markers coming from anyone, accent or not—he'd only gone to school for four years of his life, and he'd never studied science. The only kind of cell he'd heard of was the kind Zakariyya was living in out at Hagerstown. So he did what he'd always done when he didn't understand something a doctor said: he nodded and said yes.

Years later, when I asked McKusick if anyone had tried to get informed consent from the Lacks family, he said, "I suspect there was no effort to explain anything in great detail. But I don't believe anyone would have told them we were testing for cancer because that wasn't the case. They would have just said, 'Your mother had cancer, the cells from that cancer have been growing all over the place and studied in great detail, in order to understand that better, we would like to have that blood from you people.' "

When I asked Susan Hsu the same question she said, "No. We never gave consent form because you just go to draw blood. We are not doing some kind of medical research, you know, not long term. All we wanted is a few tubes of blood and to do genetic marker test. It's not involved in a human research committee or things like that."

Although this attitude wasn't uncommon at the time, NIH guidelines stipulated that all human subject research funded by NIH—as

McKusick's was—required both informed consent and approval from a Hopkins review board. Those guidelines had been implemented in 1966, in the aftermath of the Southam trial, and then expanded to include a detailed definition of informed consent in 1971. They were in the process of being codified into law when Hsu called Day.

McKusick began his research on the Lacks family at a time of great flux in research oversight. Just one year earlier, in response to Tuskegee and several other unethical studies, the Department of Health, Education, and Welfare (HEW) had launched an investigation into federal oversight of human-subject research and found it to be inadequate. As one government report said, it was a time filled with "widespread confusion about how to assess risk," as well as "refusal by some researchers to cooperate" with oversight, and "indifference by those charged with administering research and its rules at local institutions." After halting the Tuskegee study, HEW proposed new Protection of Human Subjects regulations that would require, among other things, informed consent. A notice inviting public comment on that proposed new law would be published in the *Federal Register* in October 1973, just a few months after Hsu called Day.

After Day got off the phone with Hsu, he called Lawrence, Sonny, and Deborah, saying, "You got to come over to the house tomorrow, doctors from Hopkins coming to test everybody's blood to see if you all got that cancer your mother had."

When Henrietta died, Day had agreed to let her doctors do an autopsy because they'd told him it might help his children someday. *They must have been telling the truth,* Day thought. Zakariyya was in Henrietta's womb when she first got the cancer, and he'd had all those anger problems ever since. Now Deborah was almost twenty-four, not much younger than Henrietta had been when she died. It made sense they were calling saying it was time for her to get tested.

Deborah panicked. She knew her mother had gotten sick at thirty, so she'd long feared her own thirtieth birthday, figuring that whatever happened to her mother at that age would happen to her too. And Deborah couldn't stand the idea of her own children growing up motherless like she had. At that point, LaTonya was two, Alfred was six, and Cheetah had never paid child support. Deborah had tried welfare for three months but hated it, so now she was working days at a suburban Toys "R" Us that took more than an hour and three buses to get to, then nights at a hamburger place called Gino's behind her apartment.

Since Deborah couldn't afford a babysitter, her boss at Gino's let Tonya and Alfred sit in the corner of the restaurant at night while Deborah worked. On her eight-thirty dinner break, Deborah would run behind the building to her apartment and put the children to bed. They knew not to open the door unless they heard her secret knock, and they never put the kerosene lamps near a curtain or blanket. Deborah practiced fire drills with them in case something went wrong while she was at work, teaching them to crawl to the window, throw out a sheet-rope she kept tied to the bed leg, and climb to safety.

Those children were all Deborah had, and she wasn't going to let anything happen to them. So when her father called saying Hopkins wanted to test to see if she had her mother's cancer, Deborah sobbed, saying, "Lord don't take me away from my babies, not now, not after everything we been through."

A few days after Susan Hsu's phone call, Day, Sonny, Lawrence, and Deborah all sat around Lawrence's dining room table as Hsu and a doctor from McKusick's lab collected tubes of blood from each of them.

For the next several days, Deborah called Hopkins again and again, telling the switchboard operators, "I'm calling for my cancer results." But none of the operators knew what tests she was talking about, or where to send her for help.

Soon, Hsu wrote a letter to Lawrence asking if she could send a nurse out to Hagerstown to collect samples from Zakariyya in prison. She included a copy of the George Gey tribute written by McKusick and Jones, saying she thought Lawrence would like to see an article about his mother's cells. No one in the family remembers reading that article—they figure Lawrence just put it in a drawer and forgot about it.

The Lacks men didn't think much about their mother's cells or the cancer tests. Lawrence was working full-time on the railroad and living in a house filled with children, Zakariyya was still in jail, and times had gotten tough for Sonny, who was now busy selling drugs.

But Deborah couldn't stop worrying. She was terrified that she might have cancer, and consumed with the idea that researchers had done—and were perhaps still doing—horrible things to her mother. She'd heard the stories about Hopkins snatching black people for research, and she'd read an article in *Jet* about the Tuskegee study that suggested doctors might have actually injected those men with syphilis in order to study them. "The injection of disease-causing organisms into unaware human subjects has occurred before in American medical science," the article explained. "It was done eight years ago in New York City by Dr. Chester Southam, a cancer specialist who injected live cancer cells into chronically ill elderly patients."

Deborah started wondering if instead of testing the Lacks children for cancer, McKusick and Hsu were actually injecting them with the same bad blood that had killed their mother. She started asking Day a lot of questions about Henrietta: How'd she get sick? What happened when she died? What did those doctors do to her? The answers seemed to confirm her fears: Day told her that Henrietta hadn't seemed sick at all. He said he took her into Hopkins, they started doing treatments, then her stomach turned black as coal and she died. Sadie said the same thing, and so did all the other cousins. But when she asked what kind of cancer her mother had, what treatments the doctors gave her, and what part of her was still alive, the family had no answers.

So when one of McKusick's assistants called Deborah and asked her to come into Hopkins to give more blood, she went, thinking that if her family couldn't answer questions about her mother, maybe the scientists could. She didn't know the blood was for a researcher in

California who wanted some samples for his own HeLa research, and she didn't know why McKusick's assistant was calling her and not her brothers—she figured it was because the problem her mother had didn't affect boys. She still thought she was being tested for cancer.

Deborah went into McKusick's office to give more blood on June 26, 1974, four days before the new federal law went into effect requiring Institutional Review Board (IRB) approval and informed consent for all federally funded research. The new law—published in the *Federal Register* one month earlier—applied to all "subjects at risk," meaning "any individual who may be exposed to the possibility of injury, including physical, psychological, or social injury, as a consequence of participation as a subject." But what constituted "injury" and "risk" was heavily debated. Numerous researchers had appealed to HEW, asking that collection of blood and tissues be exempt from the new law. After all, doctors had been drawing blood for centuries for diagnostic testing, and aside from the pain of a needle stick, there seemed to be no risk. But HEW did not exempt those procedures; in fact, it later clarified the law to specifically include them.

McKusick's research on the Lacks family coincided with the beginning of a new era of genetic research, in which the concept of risk to patients would change completely. With the ability to identify genes from a blood sample or even a single cell, the risk of a blood draw was no longer just a minor infection or the pain of a needle stick—it was that someone could uncover your genetic information. It was about violation of privacy.

Deborah met McKusick only once, when she went into Hopkins to give blood. He shook her hand and said that Henrietta had made an important contribution to science. Then Deborah bombarded him with questions: What made her mother sick? How was part of her still living? What did that mean? What did Henrietta do for science? And did all those blood tests he was doing mean Deborah was going to die young like her mother?

McKusick did not explain why he was having someone draw blood from Deborah. Instead he told her about Henrietta's cells being used for the polio vaccine and genetic research; he said they'd gone up in early space missions and been used in atomic bomb testing. Deborah heard those things and imagined her mother on the moon and being blown up by bombs. She was terrified and couldn't stop wondering if the parts of her mother they were using in research could actually feel the things scientists were doing to them.

When she asked McKusick to explain more about the cells, he gave her a book he'd edited called *Medical Genetics,* which would become one of the most important textbooks in the field. He said it would tell her everything she needed to know, then autographed the inside front cover. Beneath his signature he wrote a phone number and told her to use it for making appointments to give more blood.

McKusick flipped to the second page of the introduction. There, between graphs of "Disease Specific Infant Mortality" and a description of "the homozygous state of Garrodian inborn errors," was the photograph of Henrietta with her hands on her hips. He pointed to the paragraph that mentioned her:

> Parenthetically, medical geneticists making use of the study of cells in place of the whole patient have "cashed in" on a reservoir of morphologic, biochemical, and other information in cell biology derived in no small part from study of the famous cell line cultured from the patient pictured on this page, Henrietta Lacks.

The book was filled with complicated sentences explaining Henrietta's cells by saying, "its atypical histology may correlate with the unusually malignant behavior of the carcinoma," and something about the "correlate of the tumor's singularity."

Reading magazines took Deborah a long time because she had to stop often to look words up in her dictionary. Now she sat in the clinic gripping McKusick's book, not even trying to read the words.

All she could think was that she'd never seen that photograph of her mother before. *What happened to her to make her end up in there?* she wondered. *And how did he get that picture?* Day swore he'd never given it to McKusick or any of Henrietta's doctors; Deborah's brothers swore they hadn't either. The only thing Day could figure was that maybe Howard Jones had asked Henrietta for a picture, then stuck it in her medical record. But as far as Day knew, no one had ever asked permission to publish it.

When I talked to McKusick several years before his death in 2008, he was seventy-nine and still conducting research and training young scientists. He didn't recall where he'd gotten the photo, but he imagined Henrietta's family must have given it to Howard Jones or another doctor at Hopkins. Though McKusick remembered the research he conducted on the Lacks family, he didn't remember meeting Deborah or giving her his book, and said he'd never had firsthand contact with the family. He'd left that up to Hsu.

When I talked to Susan Hsu, now a director of medical genetics at the American Red Cross, she told me that working with McKusick on HeLa cells was a highlight of her career. "I'm very proud," she told me. "I probably will Xerox these paper and tell my kids this is important." But when I explained to her that the Lackses thought she was testing them for cancer, and that they were upset about scientists using the cells without their knowledge, she was shocked.

"I feel very bad," she said. "People should have told them. You know, we never thought at that time they did not understand."

She also told me she had a message she hoped I'd give to the Lacks family when I talked with them next: "Just tell them I'm really grateful," she said. "They should be very proud of the mother or the wife—I think that if they are angry probably they didn't realize how famous the cells are now in the world. It's unfortunate thing what happened, they still should be very proud, their mother will never die as long as the medical science is around, she will always be such a famous thing."

Toward the end of our conversation, Hsu mentioned that she could learn much more from testing the family's blood today, since DNA technology had advanced so much since the seventies. Then she asked if I'd tell the Lacks family one more thing for her: "If they are willing," she said, "I wouldn't mind to go back and get some more blood."

24

"Least They
Can Do"

The Lackses didn't know anything about the HeLa contamination problem that led McKusick and Hsu to them until Michael Rogers, a young reporter for *Rolling Stone,* showed up at their house with long hair and rock-and-roll clothes.

Rogers was something of a journalism prodigy. By his nineteenth birthday he'd gotten a degree in creative writing and physics and published his first story in *Esquire;* by his early twenties, when he started looking into the HeLa story, he'd already published two books and joined the staff of *Rolling Stone.* In coming years he'd go on to be an editor at *Newsweek,* and later the *Washington Post.*

Rogers first learned about HeLa cells after seeing "Helen Lane Lives!" written over a urinal in a medical school bathroom. He started reading news reports about HeLa cells and the contamination problem and realized it would make a great story for *Rolling Stone*—the perfect mix of science and human interest. So Rogers set out to find this mysterious Helen Lane.

He called Margaret Gey, who was friendly and talkative until Rogers asked about Helen Lane. Then she told him it wouldn't be a good idea for them to meet and hung up. Eventually Rogers found his way to Walter Nelson-Rees, who mentioned as an aside that Henrietta Lacks was the real name of the woman behind the cells. Soon, while sitting on his Baltimore hotel bed with the view of the B-R-O-M-O-S-E-L-T-Z-E-R clock, Rogers found Lawrence Lacks in the phone book.

It was the winter of 1975, the streets were icy, and on his way to Lawrence's house, Rogers's taxi was hit by another car in the middle of an intersection. The cab spun in the road, doing five, then six full circles, as if some giant hand had reached down and spun it like a bottle. Rogers had done risky reporting all over the world; now he was sitting in the back of a cab, gripping the door handle, thinking, *Damn it! It would be really stupid if I got killed in Baltimore working on* this *of all assignments. It's not even a dangerous story!*

Decades later, as I talked with Rogers in his Brooklyn apartment, we agreed, only half joking, that the spinning cab was probably no accident. Deborah would later say that it was Henrietta warning him to leave her family alone, because he was about to tell them something upsetting. She'd also say that Henrietta started the famous Oakland, California, fire that later burned Rogers's house, destroying all the notes and documents he'd collected about HeLa and Henrietta's family.

When Rogers made it to Lawrence's house, he expected to interview the Lackses about Henrietta, but found himself bombarded with questions instead.

"It was so clear they hadn't been treated well," Rogers told me. "They truly had *no* idea what was going on, and they really wanted to understand. But doctors just took blood samples without explaining anything and left the family worrying."

Lawrence asked, "What I was wondering was, about these cells . . . They say they're stronger, they're taking over—is that bad or good? Does that mean if we get sick, we'll live longer?"

Rogers told the Lackses that no, the cells being immortal didn't mean they'd become immortal too, or that they'd die of cancer. But he wasn't sure they believed him. He explained the concept of cells as best he could, told them about the media reports that had already appeared about HeLa, and promised he'd send them copies to read.

At that point no one in Henrietta's immediate family except Deborah seemed particularly upset about Henrietta's story or the existence of those cells.

"I didn't feel too much about the cells when I first found they was livin," Sonny told me years later. "Long as it's helpin somebody. That's what I thought."

But that changed when he and his brothers read Rogers's article and learned this:

> Cell lines are swapped, traded, forwarded, begged and borrowed among research institutions around the world. . . . The institutional sources of cells now range from [government]-supported facilities like Nelson-Rees's to commercial outfits with toll-free 800 numbers, from whom one can order, for about $25, a tiny glass vial of HeLa cells.

With that paragraph, suddenly the Lacks brothers became very interested in the story of HeLa. They also became convinced that George Gey and Johns Hopkins had stolen their mother's cells and made millions selling them.

But in fact, Gey's history indicates that he wasn't particularly interested in science for profit: in the early 1940s he'd turned down a request to create and run the first commercial cell-culture lab. Patenting cell lines is standard today, but it was unheard of in the fifties; regardless, it seems unlikely that Gey would have patented HeLa. He didn't even patent the roller drum, which is still used today and could have made him a fortune.

In the end, Gey made a comfortable salary from Hopkins, but he wasn't wealthy. He and Margaret lived in a modest home that he bought from a friend for a one-dollar down payment, then spent years fixing up and paying off. Margaret ran the Gey lab for more than a decade without pay. Sometimes she couldn't make their house payments or buy groceries because George had drained their account yet again buying lab equipment they couldn't afford. Eventually she made him open a separate checking account for the lab, and kept him away from their personal money as much as she could. On their thirtieth wedding anniversary, George gave Margaret a check for one hundred dollars, along with a note scribbled on the back of an aluminum oxide wrapper: "Next 30 years not as rough. Love, George." Margaret never cashed the check, and things never got much better.

Various spokespeople for Johns Hopkins, including at least one past university president, have issued statements to me and other journalists over the years saying that Hopkins never made a cent off HeLa cells, that George Gey gave them all away for free.

There's no record of Hopkins and Gey accepting money for HeLa cells, but many for-profit cell banks and biotech companies have. Microbiological Associates—which later became part of Invitrogen and BioWhittaker, two of the largest biotech companies in the world—got its start selling HeLa. Since Microbiological Associates was privately owned and sold many other biological products, there's no way to know how much of its revenue came specifically from HeLa. The same is true for many other companies. What we do know is that today, Invitrogen sells HeLa products that cost anywhere from $100 to nearly $10,000 per vial. A search of the U.S. Patent and Trademark Office database turns up more than seventeen thousand patents involving HeLa cells. And there's no way to quantify the professional gain many scientists have achieved with the help of HeLa.

The American Type Culture Collection—a nonprofit whose funds go mainly toward maintaining and providing pure cultures for science—has been selling HeLa since the sixties. When this book went to press, their price per vial was $256. The ATCC won't reveal how much money it brings in from HeLa sales each year, but since HeLa is one of the most popular cell lines in the world, that number is surely significant.

Lawrence and Sonny knew none of this. All they knew was that Gey had grown their mother's cells at Hopkins, someone somewhere was making money off of them, and that someone wasn't related to Henrietta Lacks. So, in an attempt to get Hopkins to give them what they saw as their cut of the HeLa profits, they made handouts about Henrietta Lacks's family being owed their due, and gave them to customers at Lawrence's store.

Deborah wanted nothing to do with fighting Hopkins—she was too busy raising her children and trying to teach herself about her mother's cells. She got herself some basic science textbooks, a good dictionary, and a journal she'd use to copy passage after passage from biology textbooks: "Cell is a minute portion of living substance," she wrote. "They create and renew all parts of the body." But mostly she wrote diary entries about what was happening:

> going on with pain
> . . . we should know what's going on with her cells from
> all of them that have her cells. You might want to ask
> why so long with this news, well its been out for years
> in and out of video's papers, books, magazines, radio, tv,
> all over the world. . . . I was in shock. Ask, and no one
> answers me. I was brought up to be quiet, no talking, just
> listen. . . . I have something to talk about now, Henrietta
> Lacks what went out of control, how my mother went
> through all that pain all by her self with those cold
> hearted doctor. Oh, how my father, said how they cooked
> her alive with radiation treatments. What went on in
> her mind in those short months. Not getting better and
> slipping away from her family. You see I am trying to
> relive that day in my mind. Youngest baby in the hospital
> with TB oldest daughter in another hospital, and three
> others at home, and husband got to, you hear me, got to
> work through it all to make sure he can feed his babies.
> And wife dying . . . Her in that cold looking ward at
> John Hopkin Hospital, the side for Black's only, oh yes,
> I know. When that day came, and my mother died, she

was Robbed of her cells and John Hopkins Hospital
learned of those cells and kept it to themselfs, and gave
them to who they wanted and even changed the name to
HeLa cell and kept it from us for 20+ years. They say
Donated. No No No Robbed Self.
My father have not signed any paper. . . . I want them to
show me proof. Where are they.

The more Deborah struggled to understand her mother's cells, the more HeLa research terrified her. When she saw a *Newsweek* article called PEOPLE-PLANTS that said scientists had crossed Henrietta Lacks's cells with tobacco cells, Deborah thought they'd created a human-plant monster that was half her mother, half tobacco. When she found out scientists had been using HeLa cells to study viruses like AIDS and Ebola, Deborah imagined her mother eternally suffering the symptoms of each disease: bone-crushing pain, bleeding eyes, suffocation. And she was horrified by reports of a "psychic healer" who, while conducting research into whether spiritual healing could cure cancer, attempted to kill HeLa cells by a laying on of hands. He wrote:

> As I held the flask, I concentrated on the picture I'd formed in my mind of the cells, visualizing a disturbance in the cell fields and the cells blowing up. . . . While I worked, I could feel a virtual tug-of-war going on between my hands and the cells' powerful adhesive ability. . . . Then I felt the field give way, as I had broken through . . . the cells looked as though someone had put a tiny hand grenade into each one—the whole culture had just blown apart! The number of dead floating cells had increased twenty times!

To Deborah, this sounded like a violent assault on her mother. But what bothered her most was the fact that so many scientists and journalists around the world continued to call her mother Helen Lane.

Since they gone ahead and taken her cells and they been so important for science, Deborah thought, *least they can do is give her credit for it.*

On March 25, 1976, when Mike Rogers's *Rolling Stone* article hit newsstands, it was the first time anyone had told the true story of Henrietta Lacks and her family, the first time the mainstream media had reported that the woman behind HeLa was black. The timing was explosive. News of the Tuskegee study was still fresh; the Black Panthers had been setting up free clinics for black people in local parks and protesting what they saw as a racist health-care system; and the racial story behind HeLa was impossible to ignore. Henrietta was a black woman born of slavery and sharecropping who fled north for prosperity, only to have her cells used as tools by white scientists without her consent. It was a story of white selling black, of black cultures "contaminating" white ones with a single cell in an era when a person with "one drop" of black blood had only recently gained the legal right to marry a white person. It was also the story of cells from an uncredited black woman becoming one of the most important tools in medicine. This was big news.

Rogers's article caught the attention of several other journalists, who contacted the Lackses. In the three months following Rogers's story, *Jet, Ebony, Smithsonian,* and various newspapers published articles about Henrietta, "one of the pivotal figures in the crusade against cancer."

Meanwhile, Victor McKusick and Susan Hsu had just published the results of their research in *Science*: in a table that took up about half of a page, under the headings "Husband," "Child 1," "Child 2," "H. Lacks," and "HeLa," McKusick, Hsu, and several coauthors mapped forty-three different genetic markers present in DNA from Day and two of the Lacks children, and used those to create a map of Henrietta's DNA that scientists could use to help identify HeLa cells in culture.

Today, no scientist would dream of publishing a person's name with any of their genetic information, because we know how much can be deduced from DNA, including the risks of developing certain diseases. Publishing personal medical information like this could violate the 1996 Health Insurance Portability and Accountability Act (HIPAA) and result in fines up to $250,000 and up to ten years in jail. It could also violate the 2008 Genetic Information Nondiscrimination Act, created to protect people from losing their health insurance or employment due to genetic discrimination. But there was no such federal oversight at the time.

A lawyer might have told the Lackses they could sue on the grounds of privacy violation or lack of informed consent. But the Lackses didn't talk to a lawyer—they didn't even know anyone had done research on their DNA, let alone published it. Deborah was still waiting to hear the results of what she thought was her cancer test, and Sonny and Lawrence were still busy trying to figure out how to get money from Hopkins. They didn't know that on the other side of the country, a white man named John Moore was about to begin fighting the same battle. Unlike the Lacks family, he knew who'd done what with his cells, and how much money they'd made. He also had the means to hire a lawyer.

25

"Who Told You You Could Sell My Spleen?"

In 1976—the same year Mike Rogers published his article in *Rolling Stone* and the Lacks family found out people were buying and selling Henrietta's cells—John Moore was working twelve-hour days, seven days a week, as a surveyor on the Alaska Pipeline. He thought the job was killing him. His gums bled, his belly swelled, bruises covered his body. It turned out that at the age of thirty-one, Moore had hairy-cell leukemia, a rare and deadly cancer that filled his spleen with malignant blood cells until it bulged like an overfilled inner tube.

Moore's local doctor referred him to David Golde, a prominent cancer researcher at UCLA, who said that removing his spleen was the only way to go. Moore signed a consent form saying the hospital could "dispose of any severed tissue or member by cremation," and Golde removed his spleen. A normal spleen weighs less than a pound; Moore's weighed twenty-two.

After the surgery, Moore moved to Seattle, became an oyster salesman, and went on with his life. But every few months between 1976 and 1983, he flew to Los Angeles for follow-up exams with

Golde. At first Moore didn't think much of the trips, but after years of flying from Seattle to L.A. so Golde could take bone marrow, blood, and semen, he started thinking, *Can't a doctor in Seattle do this?* When Moore told Golde he wanted to start doing his follow-ups closer to home, Golde offered to pay for the plane tickets and put him up in style at the Beverly Wilshire. Moore thought that was odd, but he didn't get suspicious until one day in 1983 — seven years after his surgery — when a nurse handed him a new consent form that said:

> I (do, do not) voluntarily grant to the University of California all rights I, or my heirs, may have in any cell line or any other potential product which might be developed from the blood and/or bone marrow obtained from me.

At first, Moore circled "do." Years later, he told *Discover* magazine, "You don't want to rock the boat. You think maybe this guy will cut you off, and you're going to die or something."

But Moore suspected Golde wasn't being straight with him, so when the nurse gave him an identical form during his next visit, Moore asked Golde whether any of the follow-up work he was doing had commercial value. According to Moore, Golde said no, but Moore circled "do not," just in case.

After his appointment, Moore went to his parents' house nearby. When he got there, the phone was ringing. It was Golde, who'd already called twice since Moore left the hospital. He said Moore must have accidentally circled the wrong option on the consent form, and asked him to come back and fix it.

"I didn't feel comfortable confronting him," Moore told a journalist years later, "so I said, 'Gee, Doctor, I don't know how I could have made that mistake.' But I said I couldn't come back, I had to fly to Seattle."

Soon the same form appeared in Moore's mailbox at home with a sticker that said "Circle I do." He didn't. A few weeks later he got a letter from Golde telling him to stop being a pain and sign the form.

That's when Moore sent the form to a lawyer, who found that Golde had devoted much of the seven years since Moore's surgery to developing and marketing a cell line called Mo.

Moore told another reporter, "It was very dehumanizing to be thought of as Mo, to be referred to as Mo in the medical records: 'Saw Mo today.' All of a sudden I was not the person Golde was putting his arm around, I was Mo, I was the cell line, like a piece of meat."

Weeks before giving Moore the new consent form—after years of "follow-up" appointments—Golde had filed for a patent on Moore's cells, and several extremely valuable proteins those cells produced. Golde hadn't yet sold the rights to the patent, but according to the lawsuit Moore eventually filed, Golde had entered into agreements with a biotech company that gave him stocks and financing worth more than $3.5 million to "commercially develop" and "scientifically investigate" the Mo cell line. At that point its market value was estimated to be $3 billion.

Nothing biological was considered patentable until a few years before Moore's lawsuit, in 1980, when the Supreme Court ruled on the case of Ananda Mohan Chakrabarty, a scientist working at General Electric who'd created a bacterium genetically engineered to consume oil and help clean up oil spills. He filed for a patent, which was denied on the grounds that no living organism could be considered an invention. Chakrabarty's lawyers argued that since normal bacteria don't consume oil, Chakrabarty's bacteria weren't naturally occurring—they only existed because he'd altered them using "human ingenuity."

Chakrabarty's victory opened up the possibility of patenting other living things, including genetically modified animals and cell lines, which didn't occur naturally outside the body. And patenting cell lines didn't require informing or getting permission from the "cell donors."

Scientists are quick to point out that John Moore's cells were exceptional, and few cell lines are actually worth patenting. Moore's cells produced rare proteins that pharmaceutical companies could use to treat infections and cancer. They also carried a rare virus called HTLV, a distant cousin of the HIV virus, which researchers hoped to use to create a vaccine that could stop the AIDS epidemic. Because of this, drug companies were willing to pay enormous sums to work with his cells. Had Moore known this before Golde patented them, he could have approached the companies directly and worked out a deal to sell the cells himself.

In the early 1970s a man named Ted Slavin had done precisely that with antibodies from his blood. Slavin was born a hemophiliac in the 1950s, when the only available treatment involved infusions of clotting factors from donor blood, which wasn't screened for diseases. Because of that, he'd been exposed to the hepatitis B virus again and again, though he didn't find out until decades later, when a blood test showed extremely high concentrations of hepatitis B antibodies in his blood. When the results of that blood test came back, Slavin's doctor—unlike Moore's—told him his body was producing something extremely valuable.

Researchers around the world were working to develop a vaccine for hepatitis B, and doing so required a steady supply of antibodies like Slavin's, which pharmaceutical companies were willing to pay large sums for. This was convenient, because Slavin needed money. He worked odd jobs waiting tables and doing construction, but he'd eventually have another hemophilia attack and end up unemployed again. So Slavin contacted laboratories and pharmaceutical companies to ask if they wanted to buy his antibodies. They said yes in droves.

Slavin started selling his serum for as much as ten dollars a milliliter—at up to 500 milliliters per order—to anyone who wanted it. But he wasn't just after money. He wanted someone to cure hepatitis B. So he wrote a letter to Nobel Prize–winning virologist Baruch Blumberg, who'd discovered the hepatitis B antigen and created the blood test that found Slavin's antibodies in the first place. Slavin offered Blumberg unlimited free use of his blood and tissues for his research, which began a years-long partnership. With the help of Slavin's serum, Blumberg eventually uncovered the link between hepatitis B and liver cancer, and created the first hepatitis B vaccine, saving millions of lives.

Slavin realized he probably wasn't the only patient with valuable blood, so he recruited other similarly endowed people and started a company, Essential Biologicals, which eventually merged with another, larger biological-product corporation. Slavin was only the first of many who have since turned their bodies into businesses, including nearly two million Americans who currently sell their blood plasma, many of them on a regular basis.

Moore, however, couldn't sell the Mo cells because that would have violated Golde's patent. So in 1984, Moore sued Golde and UCLA for deceiving him and using his body in research without consent; he also claimed property rights over his tissues and sued Golde for stealing them. With that, he became the first person to legally stake a claim to his own tissue and sue for profits and damages.

When Judge Joseph Wapner, most famous for being the judge on *The People's Court* television show, ended up refereeing the depositions, Moore figured no one would take the case seriously. But scientists worldwide panicked. If tissue samples—including blood cells—became patients' property, researchers taking them without getting consent and property rights up front would risk being charged with theft. The press ran story after story quoting lawyers and scientists saying that a victory for Moore would "create chaos for researchers" and "[sound] the death knell to the university physician-scientist." They called it "a threat to the sharing of tissue for research purposes," and worried that patients would block the progress of science by holding out for excessive profits, even with cells that weren't worth millions like Moore's.

But plenty of science was already on hold while researchers, universities, and biotech companies sued one another over ownership of various cell lines. Only two of those cases mentioned the people those cells came from: the first, in 1976, involved ownership of an important human-fetal-cell line. Leonard Hayflick, the researcher who'd originally grown the cells, argued that there were numerous parties with legitimate property interests in any cultured cells, including the scientist who grew them, the financers of any related work, and the "donors" of the original samples. Without any one of those contributions, he said, the cultured cells wouldn't exist, and neither would any money resulting from their sale. That case set no precedent because it settled out of court, with rights to the cells being divided between the parties involved in the lawsuit, which didn't include the cell "donor." The same was true of another case soon after, in which a young scientist took a cell line he'd helped develop in the United States and fled with it to his native Japan, claiming ownership because the original cells had come from his mother.

The public didn't realize there was big money in cell lines until news of the Moore case hit, and headlines nationwide said things like:

OWNERSHIP OF CELLS RAISES STICKY ISSUES . . .
WHO SHOULD HAVE RIGHTS TO A PATIENT'S CELLS? . . .
WHO TOLD YOU YOU COULD SELL MY SPLEEN?

Scientists, lawyers, ethicists, and policymakers debated the issues: some called for legislation that would make it illegal for doctors to take patients' cells or commercialize them without consent and the disclosure of potential profits; others argued that doing so would create a logistical nightmare that would put an end to medical progress.

Ultimately the judge threw Moore's suit out of court, saying he had no case. Ironically, in his decision, the judge cited the HeLa cell line as a precedent for what happened with the Mo cell line. The fact that no one had sued over the growth or ownership of the HeLa cell line, he said, illustrated that patients didn't mind when doctors took their cells and turned them into commercial products. The judge believed Moore was unusual in his objections. But in fact, he was simply the first to realize there was something potentially objectionable going on.

Moore appealed, and in 1988 the California Court of Appeals ruled in his favor, pointing to the Protection of Human Subjects in Medical Experimentation Act, a 1978 California statute requiring that research on humans respect the "right of individuals to determine what is done to their own bodies." The judges wrote, "A patient must have the ultimate power to control what becomes of his or her tissues. To hold otherwise would open the door to a massive invasion of human privacy and dignity in the name of medical progress."

But Golde appealed and won. And with each new decision in the suit, headlines flip-flopped:

COURT RULES CELLS ARE THE PATIENT'S PROPERTY . . .
COURT BACKS DOCTORS' RIGHT TO USE PATIENT TISSUES

Nearly seven years after Moore originally filed suit, the Supreme Court of California ruled against him in what became the definitive statement on this issue: When tissues are removed from your body, with or without your consent, any claim you might have had to owning them vanishes. When you leave tissues in a doctor's office or a lab, you abandon them as waste, and anyone can take your garbage and sell it. Since Moore had abandoned his cells, they were no longer a product of his body, the ruling said. They had been "transformed" into an invention and were now the product of Golde's "human ingenuity" and "inventive effort."

Moore wasn't awarded any of the profits, but the judge did agree with him on two counts: lack of informed consent, because Golde hadn't disclosed his financial interests, and breach of fiduciary duty, meaning Golde had taken advantage of his position as doctor and violated patient trust. The court said researchers should disclose financial interests in patient tissues, though no law required it. It also pointed out the lack of regulation and patient protections in tissue research, and called on legislators to remedy the situation. But it said that ruling in Moore's favor might "destroy the economic incentive to conduct important medical research," and that giving patients property rights in their tissues might "hinder research by restricting access to the necessary raw materials," creating a field where "with every cell sample a researcher purchases a ticket in a litigation lottery."

Scientists were triumphant, even smug. The dean of the Stanford

University School of Medicine told a reporter that as long as researchers disclosed their financial interests, patients shouldn't object to the use of their tissues. "If you did," he said, "I guess you could sit there with your ruptured appendix and negotiate."

Despite the widespread media coverage of the Moore suit, the Lacks family had no idea any of this was happening. As the debate over ownership of human tissues played out around the country, the Lacks brothers continued to tell anyone who'd listen that Johns Hopkins had stolen their mother's cells and owed them millions of dollars. And Deborah started handing out newsletters about her mother and the cells, saying, "I just want y'all read what's on this paper! And tell everybody! Bring it around. We want everybody in the world to know about my mother."

Henrietta and
David Lacks,
circa 1945.

Left: Elsie Lacks, Henrietta's older daughter, about five years before she was committed to Crownsville State Hospital, with a diagnosis of "idiocy." *Right:* Deborah Lacks at about age four.

The home-house where Henrietta was raised, a four-room log cabin in Clover, Virginia, that once served as slave quarters, 1999.

Henrietta's mother, Eliza Pleasant, died when Henrietta was four. Henrietta is buried somewhere in the clearing beside her mother's tombstone, in an unmarked grave.

Main Street in downtown Clover, Virginia, where Henrietta was raised, circa 1930s. COURTESY OF FRANCES WOLTZ

South Boston tobacco auction, circa 1920s. Henrietta and her family sold their crops at this auction house.

Sparrows Point workers cleaning a furnace by removing "slag," a toxic by-product of molten metal, sometime in the 1940s. COURTESY OF THE DUNDALK-PATAPSKO NECK HISTORICAL SOCIETY

Howard W. Jones, the gynecologist who diagnosed Henrietta's tumor, sometime in the 1950s.

George Gey, who directed the laboratory in which HeLa cells were first grown, circa 1951.

© ALAN MASON CHESNEY MEDICAL ARCHIVE

Henrietta Lacks's death certificate.

Sadie Sturdivant, Henrietta's cousin and close friend, in the early 1940s.

In 1949, labs had to make their own culture medium, a laborious process. In this picture, the man is stirring broth in a vat while the women filter the broth into smaller bottles. After HeLa, it became possible to order ready-made media by mail.

© HULTON-DEUTSCH COLLECTION/CORBIS

Margaret Gey and Minnie, a lab technician, in the Gey lab at Hopkins, circa mid-1960s.

COURTESY OF MARY KUBICEK

Mary Kubicek, the technician in the Gey lab who processed Henrietta's tumor sample and grew her cells in culture.

COURTESY OF MARY KUBICEK

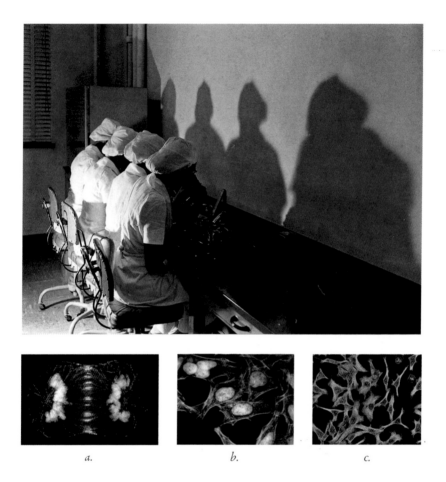

Four technicians at the Tuskegee Institute's HeLa mass production center inspecting HeLa cells before shipping them. © MARCH OF DIMES FOUNDATION

a. One HeLa cell dividing into two. COURTESY OF PAUL D. ANDREWS

b. These HeLa cells were stained with special dyes that highlight specific parts of each cell. Here, the DNA in the nucleus is yellow, the actin filaments are light blue, and the mitochondria—the cell's power generators—are pink. © OMAR QUINTERO

c. These HeLa cells were stained with fluorescent dye and photographed under a confocal microscope. COURTESY OF TOM DEERINCK

Deborah at about thirteen, the age when she was fending off her cousin Galen.

Below: Deborah with her children, LaTonya and Alfred, and her second husband, James Pullum, in the mid-1980s.

In 2001, Deborah developed a severe case of hives after learning upsetting new information about her mother and sister.

Deborah and her cousin Gary Lacks standing in front of drying tobacco, 2001.

Left: Deborah's grandson Davon, 2000. *Center:* Deborah with her brother Sonny's granddaughters, JaBrea (left) and Aiyana, 2007. *Right:* Henrietta's son Sonny with his granddaughter JaBrea, who had just been baptized, 2001.

The Lacks family in 2009. Clockwise from top right: Henrietta's middle son, Sonny (in baseball cap); Sonny's oldest daughter, Jeri; Henrietta's youngest son, Zakariyya; Henrietta's oldest son, Lawrence; Lawrence's son Ron; Deborah's grandson Alfred; Lawrence's granddaughter Courtnee; Sonny's wife, Sheryl; Sonny's son, David; Lawrence's daughter Antonetta; Sonny's son-in-law, Tom. Center: Lawrence's wife, Bobbette (in red), with her and Lawrence's granddaughter Erika (wearing glasses).

26

Breach of Privacy

Despite her fears, Deborah didn't die on her thirtieth birthday. She just kept raising her kids, working various jobs as a barber, notary public, chemical mixer at a cement plant, grocery store clerk, limousine driver.

In 1980, four years after divorcing Cheetah, Deborah took her car to a mechanic named James Pullum, who also worked at a local steel mill. They married in 1981, when Deborah was thirty-one and Pullum was forty-six, soon after he got called by the Lord to moonlight as a preacher. Pullum had some run-ins with the law before he was saved, but with him, Deborah felt safe. He rode around Baltimore on his Harley with a knife in his pocket and always had a pistol close. When he asked Deborah why he'd never met her mother, she laid the *Rolling Stone* article on the bed for him to read, and he said she should get a lawyer. She told him to mind his own business. Eventually they opened up a little storefront church, and for a while Deborah stopped worrying so much about her mother's cells.

Zakariyya was out of prison after serving only seven of his fifteen-year sentence. He'd gotten himself certified to fix air conditioners and work on trucks, but he still wrestled with anger and drinking, and on the rare occasions when he found jobs, he lost them quickly. He couldn't afford rent, so he slept most nights on a bench on Federal Hill in downtown Baltimore, or on the steps of a church across the street from his father's house. Day would sometimes look out his bedroom window and see his son lying on the concrete, but when he invited him in, Zakariyya snarled and said the ground was better. Zakariyya blamed his father for Henrietta's death, hated him for burying her in an unmarked grave, and never forgave him for leaving the children with Ethel. Day eventually stopped inviting him in, even though it sometimes meant walking past Zakariyya sleeping on the sidewalk.

At some point, Zakariyya noticed an ad seeking volunteers for medical studies at Hopkins, and he realized he could become a research subject in exchange for a little money, a few meals, sometimes even a bed to sleep on. When he needed to buy eyeglasses, he let researchers infect him with malaria to study a new drug. He volunteered for research on alcoholism to pay for a new job-training program, then signed up for an AIDS study that would have let him sleep in a bed for nearly a week. He quit when the researchers started talking about injections, because he thought they'd infect him with AIDS.

None of the doctors knew they were doing research on Henrietta Lacks's son, because he'd changed his name. Zakariyya and Deborah always figured that if Hopkins had found out he was a Lacks, they wouldn't have let him leave.

The biggest payday any of the Lacks children ever saw came when Day and other workers got a settlement from a class-action lawsuit against a boiler manufacturer over the damage done to their lungs from asbestos exposure at Bethlehem Steel. Day got a check for $12,000, and gave $2,000 to each of his children. Deborah used hers to buy a small piece of land in Clover, so she could someday move down to the country and live near her mother's grave.

Sonny's rough period was only getting worse: most of his income now came from a food-stamp ring he ran out of Lawrence's convenience store, and soon he found himself in jail for narcotics trafficking.

And it looked like Deborah's son Alfred was taking the same path as his uncles: by the age of eighteen, he'd already been arrested several times for minor offenses, like breaking and entering. After bailing him out a few times, Deborah started leaving him in jail to teach him a lesson, saying, "You just stay there till your bail come down to where you can afford it." Later, when he joined the Marines and quickly went AWOL, Deborah tracked him down and made him turn himself in to the military police. She hoped some time in minimum security would convince him he never wanted to end up in a penitentiary. But things just got worse, with Alfred stealing and coming home on drugs, and eventually Deborah realized she couldn't do a thing about it. She told him, "The devil got you, boy—that stuff you on make you crazy. I don't know you, and I don't want you around here no more."

In the midst of all this, someone told Deborah that as Henrietta's next of kin, she could request a copy of her mother's records from Hopkins to learn about her death. But Deborah didn't do it, because she was afraid of what she might find and how it might affect her. Then, in 1985, a university press published a book by Michael Gold, a reporter from *Science 85* magazine, about Walter Nelson-Rees's campaign to stop HeLa contamination. It was called *A Conspiracy of Cells: One Woman's Immortal Legacy and the Medical Scandal It Caused.*

No one in the Lacks family remembers how they learned about Gold's book, but when Deborah got a copy, she flipped through it as fast as she could, looking for her mother. She found the photo of Henrietta, hands on hips, at the front of the book, and her name at the end of the first chapter. Then she read the passage out loud to herself, shaking with excitement:

> They were all the cells of an American who in her entire life had probably not been more than a few miles from her home in Baltimore, Maryland. . . . Her name was Henrietta Lacks.

In the ten-page chapter that followed, Gold quoted extensively from her medical records: the blood spotting her underwear, the syphilis, her rapid decline. No one in Henrietta's family had ever seen those medical records, let alone given anyone at Hopkins permission to release them to a journalist for publication in a book the whole world could read. Then, without warning, Deborah turned the pages of Gold's book and stumbled on the details of her mother's demise: excruciating pain, fever, and vomiting; poisons building in her blood; a doctor writing, "Discontinue all medication and treatments except analgesics"; and the wreckage of Henrietta's body during the autopsy:

> The dead woman's arms had been pulled up and back so that the pathologist could get at her chest . . . the body had been split down the middle and opened wide . . . greyish white tumor globules . . . filled the corpse. It looked as if the inside of the body was studded with pearls. Strings of them ran over the surfaces of the liver, diaphragm, intestine, appendix, rectum, and heart. Thick clusters were heaped on top of the ovaries and fallopian tubes. The bladder area was the worst, covered by a solid mass of cancerous tissue.

After reading that passage, Deborah fell apart. She spent days and nights crying, imagining the pain Henrietta must have been in. She couldn't close her eyes without seeing her mother's body split in half, arms askew, and filled with tumors. She stopped sleeping. And soon she was as angry at Hopkins as her brothers. She stayed up nights wondering, *Who gave my mother medical records to a reporter?* Lawrence and Zakariyya thought Michael Gold must have been related to George Gey or some other doctor at Hopkins—how else could he have gotten their mother's records?

When I called Michael Gold years later, he didn't remember who'd given him the records. He said he'd had "good long conversations" with Victor McKusick and Howard Jones, and was pretty sure Jones had given him the photo of Henrietta. But he wasn't sure about the records. "They were in somebody's desk drawer," he told me. "I don't remember if it was Victor McKusick or Howard Jones." When I talked to Jones, he had no memory of Gold or his book, and denied that either he or McKusick ever gave Henrietta's medical records to anyone.

It wasn't illegal for a journalist to publish medical information given to him by a source, but doing so without contacting the subject's family to ask additional questions, verify information, and let them know such private information was being published could certainly have been considered questionable judgment. When I asked Gold whether he tried to speak to the Lacks family, he said, "I think I wrote some letters and made some calls, but the addresses and phone numbers never seemed to be current. And to be honest, the family wasn't really my focus. . . . I just thought they might make some interesting color for the scientific story."

Regardless, it was not standard practice for a doctor to hand a patient's medical records over to a reporter. Patient confidentiality has been an ethical tenet for centuries: the Hippocratic Oath, which most doctors take when graduating from medical school, says that being a physician requires the promise of confidentiality because without it, patients would never disclose the deeply personal information needed to make medical diagnoses. But like the Nuremberg Code and the American Medical Association Code of Ethics, which clearly said that doctors should keep patient information confidential, the Hippocratic Oath wasn't law.

Today, publishing medical records without permission could violate federal law. But in the early eighties, when someone gave Henrietta's medical records to Gold, there was no such law. Many states—more than thirty, in fact—had passed laws protecting the confidentiality of a patient's medical records, but Maryland was not one of them.

Several patients had successfully sued their doctors for privacy violations, including one whose medical records were released without her consent, and others whose doctors either published photographs or showed videos of them publicly, all without consent. But those patients had one thing going for them that Henrietta didn't: They were alive. And the dead have no right to privacy—even if part of them is still alive.

27

The Secret of Immortality

More than thirty years after Henrietta's death, research on HeLa cells finally helped uncover how her cancer started and why her cells never died. In 1984 a German virologist named Harald zur Hausen discovered a new strain of a sexually transmitted virus called Human Papilloma Virus 18 (HPV-18). He believed it and HPV-16, which he'd discovered a year earlier, caused cervical cancer. HeLa cells in his lab tested positive for the HPV-18 strain, but zur Hausen requested a sample of Henrietta's original biopsy from Hopkins, so he could be sure her cells hadn't been contaminated with the virus in culture. The sample didn't just test positive; it showed that Henrietta had been infected with multiple copies of HPV-18, which turned out to be one of the most virulent strains of the virus.

There are more than one hundred strains of HPV in existence, thirteen of which cause cervical, anal, oral, and penile cancer—today, around 90 percent of all sexually active adults become infected with at least one strain during their lifetimes. Throughout the eighties, using HeLa and other cells, scientists studied HPV infection and how it causes cancer. They learned that HPV inserts its DNA into the DNA of the host cell, where it produces proteins that lead to cancer. They also found that when they blocked the HPV DNA, cervical cancer cells stopped being cancerous. These discoveries would help lead to an HPV vaccine, and eventually earn zur Hausen a Nobel Prize.

Research into HPV eventually uncovered how Henrietta's cancer started: HPV inserted its DNA into the long arm of her eleventh chromosome and essentially turned off her p53 tumor suppressor gene. What scientists still haven't figured out is why this produced such monstrously virulent cells both in and out of Henrietta's body, especially since cervical cancer cells are some of the hardest of all cells to culture.

When I talked to Howard Jones fifty years after he found the tumor on Henrietta's cervix, he was in his early nineties and had seen thousands of cervical cancer cases. But when I asked if he remembered Henrietta, he laughed. "I could never forget that tumor," he said, "because it was unlike anything I've ever seen."

I talked to many scientists about HeLa, and none could explain why Henrietta's cells grew so powerfully when many others didn't even survive. Today it's possible for scientists to immortalize cells by exposing them to certain viruses or chemicals, but very few cells have become immortal on their own as Henrietta's did.

Members of Henrietta's family have their own theories about why her cells grew so powerfully: Henrietta's sister Gladys never forgave her for moving to Baltimore and leaving their father behind for Gladys to care for as he aged. The way Gladys saw it, that cancer was the Lord's way of punishing Henrietta for leaving home. Gladys's son Gary believed all disease was the wrath of the Lord—punishment for Adam eating the apple from Eve. Cootie said it was the disease-causing spirits. And Henrietta's cousin Sadie never knew what to think.

"Oh Lord," she told me once. "When I heard about them cells I thought, Could it'a been somethin live got up in her, you know? It scared me, cause we used to go around together all the time. Hennie and I ain't never been in that nasty water down there in Turners Station like the other peoples, we didn't go to no beach or nothing like that, and we didn't never go without no panties or anything, so I don't know how something got up inside Hennie. But it did. Somethin came alive up in her. She died, and it just keep on living. Made me start thinkin things, you know, like maybe something come out of space, dropped down, and she walked over it."

Sadie laughed when she said this because she knew it sounded crazy. "But that did went through my mind," she said. "I ain't lying. Everything just go through your mind, you know? How else you gonna explain them cells growin like they do?"

Every decade has had its landmark moments in HeLa research, and the connection between HPV and cervical cancer was only one of several in the eighties. At the beginning of the AIDS epidemic, a group of researchers—including a molecular biologist named Richard Axel, who would go on to win a Nobel Prize—infected HeLa cells with HIV. Normally, HIV can infect only blood cells, but Axel had inserted a specific DNA sequence from a blood cell into HeLa cells, which made it possible for HIV to infect them as well. This allowed scientists to determine what was required for HIV to infect a cell—an important step toward understanding the virus, and potentially stopping it.

Axel's research caught the attention of Jeremy Rifkin, an author and activist who was deeply involved in a growing public debate over whether scientists should alter DNA. Rifkin and many others believed that any manipulation of DNA, even in a controlled laboratory setting, was dangerous because it might lead to genetic mutations and make it possible to engineer "designer babies." Since there were no laws limiting genetic engineering, Rifkin regularly sued to stop it using any existing laws that might apply.

In 1987 he filed a lawsuit in federal court to halt Axel's research on the grounds that it violated the 1975 National Environmental Policy Act, because it had never been proven environmentally safe. It was widely known, Rifkin pointed out, that HeLa was "an extraordinarily virulent and infectious line of cells" that could contaminate other cultures. Once Axel infected HeLa cells with HIV, Rifkin said, they could infect other cells and expose lab researchers around the world to HIV, "thus increasing the virus' host range and potentially leading to the further hazardous dissemination of the AIDS virus genome."

Axel responded to the suit by explaining that cells couldn't grow outside of tissue culture and that there was a world of difference between culture contamination and HIV infection. *Science* reported on the lawsuit, writing, "Even Rifkin admits that taken together these events sound more like the plot of a grade-B horror movie than the normal run of affairs in the country's biomedical research laboratories." Eventually the suit was dismissed, Axel went on using HeLa for HIV research, and Rifkin's horror-film scenario didn't come true.

But in the meantime two scientists had developed a theory about HeLa that sounded far more like science fiction than anything Rifkin had come up with: HeLa, they said, was no longer human.

Cells change while growing in culture, just as they change in a human body. They're exposed to chemicals, sunlight, and different environments, all of which can cause DNA changes. Then they pass those changes on to each new generation of cells through cell division, a random process that produces even more changes. Like humans, they evolve.

All of this happened to Henrietta's cells once they were placed in culture. And they passed those changes on to their daughter cells, creating new families of HeLa cells that differed from one another in the same way that second, third, and fourth cousins differ, though they share a common ancestor.

By the early nineties, the little sample of Henrietta's cervix that Mary had put into culture in the Gey lab had given rise to many tons of other cells—all still known as HeLa, but all slightly different from one another, and from Henrietta. Because of this, Leigh Van Valen, an evolutionary biologist at the University of Chicago, wrote, "We here propose, in all seriousness, that [HeLa cells] have become a separate species."

Van Valen explained this idea years later, saying, "HeLa cells are evolving separately from humans, and having a separate evolution is really what a species is all about." Since the species name *Hela* was already taken by a type of crab, the researchers proposed that the new HeLa cell species should be called *Helacyton gartleri,* which combined *HeLa* with *cyton,* which is Greek for "cell," and *gartleri,* in honor of Stanley Gartler, who'd dropped the "HeLa Bomb" twenty-five years earlier.

No one challenged this idea, but no one acted on it either, so Henrietta's cells remained classified as human. But even today some scientists argue that it's factually incorrect to say that HeLa cells are related to Henrietta, since their DNA is no longer genetically identical to hers.

Robert Stevenson, one of the researchers who devoted much of his career to straightening out the HeLa contamination mess, laughed when he heard that argument. "It's just ridiculous," he told me. "Scientists don't like to think of HeLa cells as being little bits of Henrietta because it's much easier to do science when you disassociate your materials from the people they come from. But if you could get a sample from Henrietta's body today and do DNA fingerprinting on it, her DNA would match the DNA in HeLa cells."

Around the time Van Valen suggested HeLa was no longer human, researchers began exploring whether Henrietta's cells might hold the key to human life extension—perhaps even immortality—and headlines once again claimed that scientists had found the fountain of youth.

In the early 1900s, Carrel's chicken-heart cells supposedly proved that all cells had the potential for immortality. But *normal* human cells—either in culture or in the human body—can't grow indefinitely like cancer cells. They divide only a finite number of times, then stop growing and begin to die. The number of times they can divide is a specific number called the Hayflick Limit, after Leonard Hayflick, who'd published a paper in 1961 showing that normal cells reach their limit when they've doubled about fifty times.

After years of disbelief and argument from other scientists, Hayflick's paper on cell limits became one of the most widely cited in his field. It was an epiphany: scientists had been trying for decades to grow immortal cell lines using normal cells instead of malignant ones, but it had never worked. They thought their technique was the problem, when in fact it was simply that the lifespan of normal cells was preprogrammed. Only cells that had been transformed by a virus or a genetic mutation had the potential to become immortal.

Scientists knew from studying HeLa that cancer cells could divide indefinitely, and they'd speculated for years about whether cancer was caused by an error in the mechanism that made cells die when they reached their Hayflick Limit. They also knew that there was a string of DNA at the end of each chromosome called a *telomere*, which shortened a tiny bit each time a cell divided, like time ticking off a clock. As normal cells go through life, their telomeres shorten with each division until they're almost gone. Then they stop dividing and begin to die. This process correlates with the age of a person: the older we are, the shorter our telomeres, and the fewer times our cells have left to divide before they die.

By the early nineties, a scientist at Yale had used HeLa to discover that human cancer cells contain an enzyme called *telomerase* that rebuilds their telomeres. The presence of telomerase meant cells could keep regenerating their telomeres indefinitely. This explained the mechanics of HeLa's immortality: telomerase constantly rewound the ticking clock at the end of Henrietta's chromosomes so they never grew old and never died. It was this immortality, and the strength with which Henrietta's cells grew, that made it possible for HeLa to take over so many other cultures—they simply outlived and outgrew any other cells they encountered.

28

After London

The story of Henrietta Lacks eventually caught the attention of a BBC producer in London named Adam Curtis, and in 1996, he began making the documentary about Henrietta that I would later watch in Courtney Speed's beauty parlor. When Curtis arrived in Baltimore with his assistants and cameras and microphones, Deborah thought everything would change, that she and the rest of the world would learn the true story of Henrietta Lacks and the HeLa cells, and she would finally be able to move on. She started referring to periods in her life as "before London" and "after London."

Curtis and his crew covered the Lacks family story in more depth than anyone ever had, filling dozens of hours of video interviewing Deborah, prompting her from off camera to speak in complete sentences, and not wander off topic. Deborah said things like "I used to go into a corner after I was married. My husband didn't even know anything about me, you know, just being sad and crying to myself. . . . I just ask these questions in my head. . . . *Why, Lord, did you take my mother when I needed her so much?*"

The interviewer asked, "What is cancer?"

The BBC interviewed Deborah in front of the home-house in Clover; they shot Day and Sonny leaning on Henrietta's mother's tombstone, talking about what a good cook Henrietta was, and how they never heard anything about the cells until researchers called wanting blood. And they followed the Lacks family to Atlanta for a conference organized in Henrietta's honor by Roland Pattillo, the scientist who would soon steer me to Deborah.

Pattillo grew up in the thirties, the son of a blacksmith turned railroad worker in a small segregated Louisiana town. He was the first in his family to go to school, and when he learned about Henrietta as a postdoctoral fellow in Gey's lab, he felt immediately connected to her. He'd wanted to honor her contributions to science ever since. So on October 11, 1996, at Morehouse School of Medicine, he organized the first annual HeLa Cancer Control Symposium. He invited researchers from around the world to present scientific papers on cancer in minorities, and he petitioned the city of Atlanta to name October 11, the date of the conference, Henrietta Lacks Day. The city agreed and gave him an official proclamation from the mayor's office. He asked Howard Jones to contribute an article recording his memories of diagnosing Henrietta's tumor. Jones wrote:

> From a clinical point of view, Mrs. Lacks never did well. . . . As Charles Dickens said at the beginning of [A] *Tale of Two Cities*, 'It was the best of times, it was the worst of times.' But it was the best of times for science in that this very peculiar tumor gave rise to the HeLa cell line. . . . For Mrs. Lacks and the family she left behind, it was the worst of times. Scientific progress and indeed progress of all kinds is often made at great cost, such as the sacrifice made by Henrietta Lacks.

Pattillo got Deborah's phone number through a physician friend at Hopkins and called her. When she heard about his plans for the conference and the official naming of Henrietta Lacks Day, she was ecstatic: finally, a scientist was honoring her mother. Soon the Lacks family—Day, Sonny, Lawrence, Deborah, Bobbette, Zakariyya, and Deborah's grandson Davon—piled into an RV that Pattillo rented for them and drove to Atlanta, with the BBC film crew following behind.

At a gas station along the way, Deborah smiled into the camera and explained why they were headed to Morehouse.

"They gonna have a lot of doctors there talking on different subjects and different areas of the science field," she said. "And they're gonna hand out plaques to my brother and my father and me in honor of our mother name. So I know it's gonna be a great occasion."

And it was. For the first time, the Lackses were treated like celebrities: they stayed in a hotel, people asked for their autographs. But there were a few glitches. In all the excitement leading up to the ceremony, Sonny's blood pressure shot up dangerously high and he ended up in the hospital, nearly missing the whole event. Zakariyya emptied the minibar in his room, then emptied his father's and Deborah's. He yelled and threw programs when he saw that they listed him as "Joseph Lacks" and Henrietta as the woman who'd "donated" the HeLa cells.

Deborah did her best to ignore all that. When she walked onto the stage, she was so nervous the podium shook when she touched it. She'd worried for weeks that there might be a sniper in the audience—a scientist who'd want to take her out to do research on her body, or to keep the family from causing problems. But Pattillo assured her she was safe.

"Excuse me if I mispronouncing a word," she told everyone at the conference, "but I have problems and I didn't get the right teaching when I was coming up in school. I was not even allowed to have hearing aid until after I was grown. But I'm not ashamed of it."

Then, with Pattillo cheering nearby, Deborah cleared her throat and began her speech:

When Dr. Pattillo called me, it all became real. For years, it seem to be a dream. Not knowing what was going on all these years. Didn't know how to even talk about it. Can this about our mother be true? Not knowing who to go to for understanding. No one from the medical field took the time.

Then, without so much as a pause, she began talking directly to her mother:

> We miss you, Mama. . . . I think of you all the time and wish I could see and hold you in my arms, like I know you held me. My father said that you told him on your dying bed to take care of Deborah. Thank you, Ma, we will see you again someday. We read what we can and try to understand. My mind often wonder how things might would be if God had you stay here with me. . . . I keep with me all I know about you deep in my soul, because I am part of you, and you are me. We love you, Mama.

It seemed like things were going better for the Lackses, like Henrietta would finally begin getting the recognition Deborah hoped for.

Soon the BBC showed up in Turner Station, asking locals about life there in the forties and fifties. News of their visit, like news of everything else that happens in Turner Station, quickly found its way to Speed's Grocery, where Courtney Speed learned the story of Henrietta Lacks for the first time. It felt like serendipity—she and several other women had recently founded the Turner Station Heritage Committee, and they were organizing events to bring attention to black people from Turner Station who'd contributed good things to the world: a former congressman who became president of the NAACP, an astronaut, and the man who'd won several Emmy awards as the voice of Sesame Street's Elmo.

When they learned about Henrietta and HeLa, Speed and a sociologist at Morgan State University named Barbara Wyche went into overdrive. They wrote letters to Congress and the mayor's office demanding recognition of Henrietta's contribution to science. They also got in touch with Terry Sharrer, a curator at the Smithsonian National Museum of American History, who invited the Lacks family to a small event at the museum. There Day admired old farm equipment and insisted that he wanted to see his wife's cells. (The museum had a flask of HeLa in storage somewhere, the medium as dark as a murky pond, but it wasn't on display.) People came up to Deborah with tears in their eyes and told her that her mother's cells had helped them over-come cancer. Deborah was thrilled. After hearing a researcher talk about cloning, Deborah asked Sharrer whether it was possible to take DNA from HeLa cells and put it into one of Deborah's eggs to bring her mother back to life. Sharrer said no.

After the event, Sharrer wrote a letter to Wyche suggesting that, to commemorate Henrietta, she and Speed consider starting an African-American health museum in Turner Station. The women soon founded the Henrietta Lacks Health History Museum Foundation, Inc., with Speed as president. They planned events featuring Henrietta Lacks look-alikes—a few Turner Station women who'd style their hair like Henrietta's and wear suits identical to the one she wore in her iconic photo. To raise awareness of Henrietta's contribution, Speed used her own money to make and give away Henrietta Lacks T-shirts, and someone else made Henrietta Lacks pens. The local papers wrote about their plan for a $7 million museum, and Speed and Wyche opened a Henrietta Lacks Foundation bank account, filed for a tax ID number, and began trying to collect as much money and information as they could for the museum. One of their first goals was getting a life-sized wax Henrietta statue.

Deborah wasn't appointed as an officer or foundation board mem-ber, but Speed and Wyche called occasionally to ask if she'd speak at various celebrations honoring her mother—once under a small tent near Speed's Grocery, other times at a nearby church. Eventually some-one suggested that Deborah donate Henrietta's Bible and the locks of hair from Henrietta and Elsie that she kept tucked inside. It was for safekeeping, people said, in case Deborah's house ever caught fire. When Deborah heard that, she ran home and hid her mother's Bible, telling her husband, "That's the only things I have from my mother, now they want to take it!"

When she found out that Speed and Wyche had started a foundation and bank account in her mother's name, Deborah was furious. "The family don't need no museum, and they definitely don't need no wax Henrietta," she said. "If anybody collecting money for anything, it should be Henrietta children collecting money for going to the doctor."

Deborah only agreed to help with the museum project when it looked like Speed and Wyche might turn up information about her mother. The three of them hung handwritten flyers in Speed's grocery store and around Turner Station, asking, "Who knew her favorite hymn? Who knew her favorite scripture? Who knew her favorite color? Who knew her favorite game?" The first two questions were Speed's; the second two came from Deborah.

At one point Speed and Wyche invited Gey's former assistant, Mary Kubicek, to an event in the basement of the New Shiloh Baptist Church in Turner Station, to talk about how she grew HeLa cells. Mary stood wrapped in scarves on a small platform stage, nervous and going blind, as distant Lacks cousins and locals not related to Henrietta yelled questions from the audience, demanding to know who made money off the cells, and whether Gey had patented them.

"Oh no," Mary said, shifting from foot to foot. "No, no, no . . . there was no way to patent cells then." She told them that in the fifties, no one imagined such a thing might someday be possible. Gey just gave the cells away for free, she said, for the good of science.

People in the room grumbled, and tension grew. One woman stood up and said, "Them cells cured me of my cancer, if I got cells that can help somebody like her cells help me, I say take em!" Another woman said she still believed Gey had patented the cells, then yelled, "I hope in the future this could be rectified!" Deborah just fluttered around the room saying that her mother had cured cancer and everyone should just calm down. Then she asked Mary to tell the story about seeing her mother's red toenails during the autopsy—the one Deborah had read in Gold's book. Mary did, and the audience fell silent.

While Speed worked with other Turner Station residents to gather memories of Henrietta, Wyche wrote letter after letter, trying to get recognition for Henrietta and attract donors to pay for the museum. And she got results: the Maryland State Senate sent a resolution on fancy paper, saying, "Be it hereby known to all that The Senate of Maryland offers its sincerest congratulations to Henrietta Lacks." On June 4, 1997, Representative Robert Ehrlich Jr. spoke before the U.S. House of Representatives, saying, "Mr. Speaker, I rise today to pay tribute to Henrietta Lacks." He told Congress her story, saying, "Ms. Lacks was not acknowledged as the donor of the cells." He said it was time for that to change. This, everyone seemed to believe, was where Hopkins should come in.

Wyche had been working on that: she'd written a meticulously detailed three-page, single-spaced letter to William Brody, then president of Johns Hopkins. She called Henrietta an "unsung local heroine," explaining the importance of the HeLa cells, and quoting a historian saying the HeLa story was "one of the most dramatic and important in the history of research at the Johns Hopkins Medical Institution." She also wrote this:

> The [Lacks] family has suffered greatly. . . . This family is, like so many others today, attempting to grapple with the many questions and the moral and ethical issues that surround the "birth" of HeLa, and the "death" of Mrs. Lacks. . . . The questions of (1) whether or not permission was received from the "donor" or her family for either the "use" of HeLa worldwide or the "mass," and commercial, production, distribution, and marketing of Mrs. Lacks' cells . . . (2) whether or not scientists, university and government personnel and others have acted ethically in these two areas or in their interactions with the family . . . Other social issues also arise because Mrs. Lacks was an African American Woman.

One month later, Ross Jones, assistant to the president of Hopkins, replied. He said he was "uncertain what role Hopkins might play in any plan to celebrate Mrs. Lacks' life," but that he wanted to share this information with Wyche:

> Please let me emphasize that Hopkins never used the HeLa cells in a commercial venture. Hopkins never sought, nor realized, any money from the development, distribution or use of the HeLa cell cultures. In keeping with almost universally accepted practice at the time, physicians and other scientists at Hopkins and elsewhere did not seek permission to use tissue removed as part of diagnostic and treatment procedures. Also, in keeping with traditions of academic research at the time, the cultures were shared freely, without compensation and in good faith with scientists around the world who requested them. Indeed, willingness of Hopkins scientists to provide access to the cultures is perhaps the principal reason for the great benefits that have derived from their use.
>
> As I'm sure we both know, many standards of practice in academic medicine have changed dramatically in recent years, and I hope and trust that there is increased sensitivity to, and awareness of, the wishes and interests of patients when they seek medical care or participate in research. That is all to the good, for academic medicine and those we serve.

He also told Wyche that he had circulated her letter to "others at Hopkins for comment and consideration." Soon a small group of people at Hopkins began meeting unofficially, without telling Wyche or Speed, to discuss what the university might do to honor Henrietta and the Lacks family.

Then they heard about Cofield.

Sir Lord Keenan Kester Cofield was the cousin of Deborah's husband's former stepdaughter, or something like that. No one in the family remembers for sure. They also don't know how or when he learned about Henrietta's cells. What they do remember is that one day Cofield called Deborah, saying he was a lawyer and that she needed to protect herself and her mother by copyrighting the name Henrietta Lacks. He also said he believed Hopkins was guilty of medical malpractice, and that it was time to sue for the family's cut of all the money Henrietta's cells had earned since the fifties, a percentage of which he would take as his fee. He would charge nothing up front, and the Lackses wouldn't have to pay if he didn't win.

Deborah had never heard about needing to copyright anything, but the family had always thought they should talk to a lawyer about the cells, and Cofield sounded like one they could afford. Deborah's brothers were thrilled, and she soon introduced Cofield to Speed and Wyche as the family's lawyer.

Cofield began spending his days at Hopkins, digging through the medical school's archives, taking notes. Of all the people who'd come to the Lackses over the years talking about the cells, he was the first to tell the family anything specific about what happened to Henrietta at Hopkins. The way the Lackses remember it, his findings confirmed their worst fears. He told them that one of the doctors who treated Henrietta didn't have a medical license, and that another had been expelled from the American Medical Association. On top of that, Cofield said, Henrietta's doctors had misdiagnosed her cancer and might have killed her with an overdose of radiation.

He told Deborah he needed to read her mother's medical records to investigate how the doctors had treated her, and to document any possible malpractice. Since only Henrietta's family members were authorized to request her records, Deborah agreed to go with him to Hopkins, where she filled out a request form. But the photocopy machine was broken, so the woman behind the desk told Deborah and Cofield they'd have to come back later, once the machine was fixed.

When Cofield returned alone, the staff refused to give him the records because he wasn't a doctor or a relative of the patient. When Cofield said he was Dr. Sir Lord Keenan Kester Cofield, the Hopkins medical records staff contacted Richard Kidwell, one of Hopkins's attorneys. Kidwell got suspicious the moment he heard that someone was poking around Hopkins using the title "Dr. Sir Lord," so he did some quick background research.

Keenan Kester Cofield wasn't a doctor or lawyer at all. In fact, Cofield had served years in various prisons for fraud, much of it involving bad checks, and he'd spent his jail time taking law courses and launching what one judge called "frivolous" lawsuits. Cofield sued guards and state officials connected to the prisons he'd been in, and was accused of calling the governor of Alabama from jail and threatening to murder him. Cofield sued McDonald's and Burger King for contaminating his body by cooking fries in pork fat, and he threatened to sue several restaurants for food poisoning—including the Four Seasons in New York City—all while he was incarcerated and unable to eat at any restaurants. He sued The Coca-Cola Company, claiming a bottle of soda he'd bought was filled with ground glass, though he was in a prison that only offered Pepsi products in aluminum cans. He'd also been convicted of fraud for a scam in which he got an obituary of himself published, then sued the newspaper for libel and damages up to $100 million. He told the FBI that he'd filed at least 150 similar lawsuits.

In various court documents, judges described Cofield as a "con artist," "no more than a gadfly and an exploiter of the court system," and "the most litigious inmate in the system." By the time Cofield contacted the Lackses about suing Hopkins, he'd been banned from filing lawsuits in at least two counties.

But Deborah knew none of this. Cofield called himself doctor and lawyer, and seemed capable of getting and understanding more information from Hopkins than the family ever could. And his demeanor didn't hurt. When Courtney Speed described him to me a few years later, she said, "Charisma! Woo! I mean, cream of the smooth! Very well versed and knew something about everything."

When Kidwell learned the truth about Cofield, the first thing he did was protect Deborah—something the Lacks family never would have expected from someone at Hopkins. He told her that Cofield was a con artist, and had her sign a document forbidding Cofield access to her family's records. The way everyone I talked to at Hopkins remembers it, when Cofield came back and learned that the family had denied him access, he yelled and demanded copies of the records until a security guard threatened to physically remove him and call the police.

Cofield then filed a lawsuit against Deborah, Lawrence, Courtney Speed, the Henrietta Lacks Health History Museum Foundation, and a long list of Hopkins officials: the president, the medical records administrator, an archivist, Richard Kidwell, and Grover Hutchins, the director of autopsy services. He sued ten defendants in all, and several of the Hopkins employees involved had never heard of Cofield or Henrietta Lacks before their subpoenas arrived.

Cofield accused Deborah, Speed, and the museum foundation of breach of contract for entering into an agreement that required him to have access to Henrietta's medical records, then denying him access. He claimed that Deborah could not legally prohibit him from doing research for the Henrietta Lacks Health History Museum Foundation, because she was not a member of its board of directors, or officially involved with the foundation in any way. He also claimed racial discrimination, saying he was "harassed by negro security of Johns Hopkins, and staff at the archives," and that "the defendants and employees actions were all racially motivated and very anti-black." He demanded access to the medical records and autopsy reports of Henrietta and Deborah's sister, Elsie, as well as damages of $15,000 per defendant, plus interest.

The most astonishing detail of Cofield's suit was his claim that the Lacks family had no right to any information about Henrietta Lacks because she'd been born Loretta Pleasant. Since there was no official record of a name change, Cofield argued, Henrietta Pleasant had never actually existed, and therefore neither had Henrietta Lacks. Whoever she was, he said, the family wasn't legally related to her. In a statement so filled with grammatical errors it's difficult to understand, Cofield called this an "obvious fraud and conspiracy" and claimed that his lawsuit would "ultimately lead to the ends of justice for only Mrs. Henrietta Lacks, and now the plaintiff who has become the victim of a small, but big time fraud."

Piles of legal documents began arriving almost daily at Deborah's door: summonses and petitions and updates and motions. She panicked. She went to Turner Station and burst into Speed's grocery store screaming, demanding that Speed give her everything she'd gathered related to Henrietta: the documents Speed kept in a superhero pillowcase, the Henrietta Lacks T-shirts and pens, the video of Wyche interviewing Day in Speed's beauty parlor. Deborah yelled at Speed, accused her of conspiring with Cofield, and said she was going to hire O. J. Simpson's lawyer, Johnnie Cochran, and sue Speed for everything she had if she didn't shut down the foundation and stop all Henrietta-related activities.

But Speed had nothing and was just as scared as Deborah. She was a single mother with six sons, and she planned to put all of them through college using money she made cutting hair and selling chips, candy, and cigarettes. Her store was being robbed regularly, and she was getting just as many court mailings from Cofield as Deborah was. Soon, Speed stopped opening the letters and let them pile up in the backroom of her store until they stacked thirty envelopes high. Then she started a new pile. She prayed to God for the letters to stop, and wished her husband was still alive to deal with Cofield.

By this time the BBC documentary had aired, and reporters were calling Deborah, requesting photos of Henrietta and the family, and asking questions about her mother and how she died. But Deborah still didn't know anything beyond what she'd read in Gold's book. It was time, she decided, to find out what her mother's medical records said. So she requested a copy from Hopkins, along with a copy of her sister's records.

She also met with Kidwell, who told her not to worry and promised that Hopkins would fight Cofield. And it did. The case was eventually dismissed, but everyone involved was spooked. When the group at Hopkins that had been working on a plan to honor Henrietta heard about Cofield's lawsuit, they quietly dropped the idea, never telling the Lackses they'd even considered it.

Years later, when I talked to Grover Hutchins, the pathologist listed in Cofield's lawsuit, he shook his head and said, "The whole thing was very sad. They wanted to have some kind of recognition for Henrietta, but then things got so hairy with Cofield and the crazy things he was saying the family thought about Hopkins, they decided it was best to let sleeping dogs lie and not get involved with anything having to do with the Lackses."

When I talked with Johns Hopkins spokesperson JoAnn Rodgers, she said there had never been an official effort by Hopkins to honor Henrietta. "It was an individual effort—maybe one or two people—and when they went away, it went away. It was never an institutional initiative."

Though the subpoenas had finally stopped coming, Deborah didn't believe the lawsuit was truly over. She couldn't shake the idea that Cofield might send people to her house to steal her mother's Bible or the lock of hair she kept tucked inside it. Or maybe he'd try to steal her cells, thinking they might be valuable like her mother's.

She stopped checking her mail and rarely left the house except to work her shifts driving a school bus for disabled children. Then she was in a freak accident: a teenager on the bus attacked her, throwing himself on top of her, biting and scratching until two men ran onto the bus and pulled him off. A few days later the same boy attacked her again, this time permanently damaging several discs in her spine.

Deborah had her husband hang dark curtains on their windows and stopped answering her phone. Then, sitting in her dark living room a year and a half after Cofield's lawsuit ended, she finally began reading and rereading the full details of her mother's death in her medical records. And for the first time, she learned that her sister had been committed to a mental institution called Crownsville.

She began worrying that something bad had happened to her sister in that hospital. *Maybe she was used in some kind of research like our mother,* she thought. Deborah called Crownsville for a copy of Elsie's records, but an administrator said most of Crownsville's documents from before 1955, the year Elsie died, had been destroyed. Deborah immediately suspected that Crownsville was hiding information about her sister, just as she still believed Hopkins was hiding information about Henrietta.

Within hours of her call to Crownsville, Deborah became disoriented and had trouble breathing. Then she broke out in hives—red welts covering her face, neck, and body, even the soles of her feet. When she checked herself into a hospital, saying, "Everything going on with my mother and sister is making my nerves break down," her doctor said her blood pressure was so high she'd nearly had a stroke.

A few weeks after Deborah came home from the hospital, Roland Pattillo left a message on her answering machine saying he'd been talking to a reporter who wanted to write a book about Henrietta and her cells, and he thought Deborah should talk to her. That reporter was me.

29

A Village of Henriettas

For nearly a year after our first conversation, Deborah refused to talk to me. I traveled back and forth to Clover, sitting on porches and walking the tobacco fields with Cliff, Cootie, and Gladys's son Gary. I dug through archives, church basements, and the abandoned, falling-down building where Henrietta went to school. While I was on the road, I'd leave messages for Deborah every few days, hoping to convince her that if she talked to me, we could learn about Henrietta together.

"Hey, I'm in your mother's tobacco field by the home-house," I told her. "I'm on the porch with Cousin Cliff, he says hi." "I found your mother's baptism records today." "Aunt Gladys is doing well after her stroke. She told me some great stories about your mom." I imagined Deborah leaning over her answering machine listening, dying to know what I'd found.

But she never picked up.

One day her husband, the Reverend James Pullum, answered the phone on the second ring and started yelling without saying hello: "They want to be assured that they going to get some MONETARY

SATISFACTION. And until anybody makes an AGREEMENT or puts that on PAPER, they are NOT going to talk ANYMORE. Everybody's received some compensation but them, and that was they MOTHER. They just feel wrong about it. It's been a real long haul for my wife, and she really takes a trip on it. Used to be she just wanted John Hopkin to give her mother some credit and explain that cell stuff to where she understand what happened to her mother. But they ignored us, so now we just mad." Then he hung up on me.

A few days later, ten months after our first conversation, Deborah called me. When I answered the phone, she yelled, "Fine, I'll talk to you!" She didn't say who she was and didn't need to. "If I'm gonna do this, you got to promise me some things," she said. "First, if my mother is so famous in science history, you got to tell everybody to get her name right. She ain't no Helen Lane. And second, everybody always say Henrietta Lacks had four children. That ain't right, she had five children. My sister died and there's no leavin her out of the book. I know you gotta tell *all* the Lacks story and there'll be good and bad in that cause of my brothers. You gonna learn all that, I don't care. The thing I care about is, you gotta find out what happened to my mother and my sister, cause I need to know."

She took a deep breath, then laughed.

"Get ready, girl," she said. "You got no idea what you gettin yourself into."

Deborah and I met on July 9, 2000, at a bed-and-breakfast on a cobblestone street corner near the harbor in Baltimore, in a neighborhood called Fell's Point. When she saw me standing in the lobby waiting for her, she pointed to her hair and said, "See this? I'm the gray child cause I'm the one doing all the worrying about our mother. That's why I wouldn't talk to you this last year. I swore I was never talkin to nobody about my mother again." She sighed. "But here I am . . . I hope I don't regret this."

Deborah was a substantial woman—about five feet tall and two hundred pounds. Her tight curls were less than an inch long and jet black, except for a thin streak of natural gray framing her face like a headband. She was fifty, but seemed both a decade older and younger at the same time. Her smooth light brown skin was dotted with big freckles and dimples, her eyes light and mischievous. She wore capri pants and Keds sneakers and moved slowly, leaning most of her weight on an aluminum cane.

She followed me to my room, where a large flat package covered in bright, flowered wrapping paper lay on the bed. I told her it was a gift for her from a young Hopkins cancer researcher named Christoph Lengauer. He'd e-mailed me a few months earlier in response to an article I'd published in *Johns Hopkins Magazine* after meeting the Lacks men. "I felt somehow bad for the Lacks family," Lengauer wrote. "They deserved better."

He'd been working with HeLa cells daily his whole career, he said, and now he couldn't get the story of Henrietta and her family out of his mind. As a Ph.D. student, he'd used HeLa to help develop something called *fluorescence in situ hybridization,* otherwise known as FISH, a technique for painting chromosomes with multicolored fluorescent dyes that shine bright under ultraviolet light. To the trained eye, FISH can uncover detailed information about a person's DNA. To the untrained eye, it simply creates a beautiful mosaic of colored chromosomes.

Christoph had framed a fourteen-by-twenty-inch print of Henrietta's chromosomes that he'd "painted" using FISH. It looked like a photograph of a night sky filled with multicolored fireflies glowing red, blue, yellow, green, purple, and turquoise.

"I want to tell them a little what HeLa means to me as a young cancer researcher, and how grateful I am for their donation years ago," he wrote. "I do not represent Hopkins, but I am part of it. In a way I might even want to apologize."

Deborah threw her black canvas tote bag onto the floor, tore the wrapping paper from the photo, then held the frame at arm's length in front of her. She said nothing, just ran through a set of French doors onto a small patio to see the picture in the setting sunlight.

"They're beautiful!" she yelled from the porch. "I never knew they were so pretty!" She walked back inside clutching the picture, her cheeks flushed. "You know what's weird? The world got more pictures of my mother cells than it do of her. I guess that's why nobody knows who she is. Only thing left of her is them cells."

She sat down on the bed and said, "I want to go to research labs and seminars to learn what my mother cells did, talk to people that been cured of cancer." She started bouncing, excited like a little girl. "Just thinkin about that make me want to get back out there. But something always happens and I go back into hiding."

I told her Lengauer wanted her to come into his lab. "He wants to say thank you and show you your mother's cells in person."

Deborah traced her mother's chromosomes in the picture with her finger. "I do want to go see them cells, but I'm not ready yet," she said. "My father and my brothers should go too, but they think I'm crazy just comin down here. They always yellin about 'Them white folks gettin rich off our mother while we got nothin.'" Deborah sighed. "We ain't gonna get rich about any of this stuff on my mother cells. She out there helpin people in medicine and that's good, I just want the history to come out to where people know my mother, HeLa, was Henrietta Lacks. And I would like to find some information about my mother. I'm quite sure she breastfed me, but I never knew for sure. People won't talk about my mother or my sister. It's like the two of them never born."

Deborah grabbed her bag off the floor, and dumped its contents onto the bed. "This is what I got about my mother," she said, pointing to a pile on the bed. There were hours of unedited videotapes from the BBC documentary, a tattered English dictionary, a diary, a genetics textbook, many scientific journal articles, patent records, and unsent greeting cards, including several birthday cards she'd bought for Henrietta, and a Mother's Day card, which she grabbed off the pile.

"I carried this around in my purse for a long time," she said, handing it to me. The outside was white with pink flowers, and inside, in flowing script, the card said, "May the spirit of our Lord and savior be with you on this day on which you are honored for all the love you have given to your family and loved ones. With prayers and love. Happy Mother's Day." It was signed "Love, Deborah."

But mostly her bag was filled with ragged newspaper and magazine articles. She held up a story about her mother from the *Weekly World News* tabloid. It was headlined THE IMMORTAL WOMAN! and it ran between an article about a telepathic dog and another about a half-human, half-alligator child.

"When I saw this thing in the grocery store, it scared me half to death," Deborah told me. "I was like, what crazy thing they sayin happened to my mother now? Everybody always say Hopkins took black folks and experiment on them in the basement over there. Nobody could prove it so I never did believe it really. But when I found out about my mother cells, I didn't know what to think except maybe all that stuff about them experimentin on people is true."

Just a few weeks earlier, Deborah told me, Day's new wife, Margaret, came home from a doctor's appointment screaming about something she'd seen in the basement at Hopkins. "She hit some wrong button on the elevator and it took her all the way down in the basement where it was dark," Deborah told me. "The door opened up and she looked straight ahead and saw all these cages. She started yellin, 'Dale, you not gonna believe it, but them cages was filled with man-sized rabbits!' "

Deborah laughed as she told me the story. "I didn't believe it. I was like, 'Man-sized rabbits?! You crazy!' I mean, who ever heard of a man-sized rabbit? But Margaret usually honest with me, so I know she saw something got her all scared. I guess anything possible."

Then, as though she was saying something as everyday as *It's supposed to rain tomorrow,* she said, "Scientists do all kinds of experiments and you never know what they doin. I still wonder how many people they got in London walkin around look just like my mother."

"What?" I said. "Why would there be women in London who look like your mother?"

"They did that cloning on my mother over there," she said, surprised I hadn't come across that fact in my research. "A reporter came here from England talking about they cloned a sheep. Now they got stuff about cloning my mother all over." She held up an article from *The Independent* in London and pointed at a circled paragraph: "Henrietta Lacks's cells thrived. In weight, they now far surpassed the person of their origin and there would probably be more than sufficient to populate a village of Henriettas." The writer joked that Henrietta should have put ten dollars in the bank in 1951, because if she had, her clones would be rich now.

Deborah raised her eyebrows at me like, *See? I told you!*

I started saying it was just Henrietta's *cells* scientists had cloned, not Henrietta herself. But Deborah waved her hand in my face, shushing me like I was talking nonsense, then fished a videocassette from the pile and held it up for me to see. It said *Jurassic Park* on the spine.

"I saw this movie a bunch of times," she said. "They talking about the genes and taking them from cells to bring that dinosaur back to life and I'm like, *Oh Lord, I got a paper on how they were doin that with my mother's cells too!*" She held up another videocassette, this one a made-for-TV movie called *The Clone*. In it, an infertility doctor secretly harvests extra embryos from one of his patients and uses them to create a colony of clones of the woman's son, who died young in an accident.

"That doctor took cells from that woman and made them into little boys look just like her child," Deborah told me. "That poor woman didn't even know about all the clones until she saw one walk out of a store. I don't know what I'd do if I saw one of my mother clones walkin around somewhere."

Deborah realized these movies were fiction, but for her the line between sci-fi and reality had blurred years earlier, when her father got that first call saying Henrietta's cells were still alive. Deborah knew her mother's cells had grown like the Blob until there were so many of them they could wrap around the Earth several times. It sounded crazy, but it was true.

"You just never know," Deborah said, fishing two more articles from the pile and handing them to me. One was called HUMAN, PLANT CELLS FUSED: WALKING CARROTS NEXT? The other was MAN-ANIMAL CELLS BRED IN LAB. Both were about her mother's cells, and neither was science fiction.

"I don't know what they did," Deborah said, "but it all sound like *Jurassic Park* to me."

For the next three days, Deborah came to my B&B room each morning, sat on the bed, and unloaded her mind. When we needed a change of scenery, we rode water taxis and walked along the Baltimore Harbor. We ate crabs and burgers and fries and drove the city streets. We visited the houses she'd lived in as a child, most now boarded up with CONDEMNED signs out front. We spent day and night together as I soaked up as much of her story as I could, constantly worried she'd change her mind and stop talking to me. But in reality, it seemed now that Deborah had started talking, she might never stop again.

Deborah's was a world without silence. She yelled, punctuated most sentences with a raspy, high-pitched laugh, and maintained a running commentary on everything around her: "Look at the size of those trees!" "Isn't that car a nice green?" "Oh my god, I've never seen such pretty flowers." She walked down the street talking to tourists, sanitation workers, and homeless people, waving her cane at every person she passed, saying, "Hi there, how y'all doin?" again and again.

Deborah was full of oddly charming quirks. She carried a bottle of Lysol in her car that she would often spray at random, only half-joking. She sprayed directly in front of my nose several times when I sneezed, but mostly she sprayed it out the window when we stopped somewhere that looked particularly unsanitary, which happened often. She also gestured with her cane as she spoke, often tapping my shoulder with it to get my attention, or smacking it against my leg to emphasize a point.

One of the first times she hit me with her cane, we were sitting in my room. She'd just handed me a copy of *Medical Genetics*, by Victor McKusick, and said, "I met this man cause he wanted some blood from me for some cancer tests."

I told her he'd taken the blood for research on Henrietta's cells, not to test her and her brothers for cancer. That's when she smacked me on the leg with her cane.

"Dang!" she yelled. "Now you tell me! When I started asking him questions about them tests and my mother's cells, he just handed me a copy of this book, patted me on my back, and send me home." She reached over, flipped the book open, and pointed. "He autographed it for me," she said, rolling her eyes. "Would have been nice if he'd told me what the damn thing said too."

Deborah and I sprawled across the bed for hours each day, reading her files and talking about her life. Then, toward the end of the third day, I noticed a thick manila folder on my pillow.

"Are those your mother's medical records?" I asked, reaching for it.

"No!" Deborah screamed, wild-eyed, leaping up and diving onto the folder like it was a fumbled football, hugging it to her chest, curling her body around it.

I sat stunned, hand still reaching toward the pillow where the envelope had been, stammering, "I . . . I mean . . . I wasn't . . ."

"That's *right* you wasn't!" Deborah snapped. "What were you gonna do to my mother medical records?!"

"I thought you put them there for me . . . I'm sorry . . . I don't need to read them now. . . . It's fine."

"We ain't ready for that!" Deborah snapped, her eyes wide and panicked. She grabbed her bag, stuffed all her things back inside it, then ran for the door.

I was stunned. The woman I'd been lying next to for days—

laughing, elbowing, consoling—was now running from me like I was out to get her.

"Deborah!" I called after her. "I'm not trying to do anything bad. I just want to learn your mother's story, same as you."

She whipped around, her eyes still panicked, "I don't know who to trust," she hissed, then ran out the door, slamming it behind her.

30

Zakariyya

The next day, Deborah called my room from the front desk as if nothing had happened. "Come on downstairs," she said. "It's time you went and talked to Zakariyya. He been askin about you."

I was not excited to meet Zakariyya. I'd heard several times that of all the Lackses, he was by far the angriest about what happened to his mother, and that he was looking for any revenge he could get. I hoped to see the age of thirty, and it seemed like being the first white person to show up at Zakariyya's apartment asking questions about his mother might interfere with that.

Outside, as I followed Deborah to her car, she said, "Things never went quite right with Zakariyya after he got out of jail. But don't worry. I'm pretty sure he's ready to talk about our mother again."

"You're *pretty* sure?" I said.

"Well, I used to make copies of information about our mother and give it to him, but he got enough to where one day he cuss me out. He ran at me screamin, 'I don't wanna hear no more stuff about my mother and that damn doctor who done raped her cells!' We haven't really talked about it since." She shrugged. "But he says he's okay

with you askin questions today though. We just got to catch him before he start drinkin."

When we got to Deborah's car, her two grandsons—Davon and
Alfred, who were just shy of their eighth and fourth birthdays—sat in
the backseat screaming at each other. "Them are my two little hearts,"
Deborah said. They were strikingly beautiful children, with huge smiles
and wide, dark eyes. Alfred sat in the back wearing two pairs of jet-
black plastic sunglasses, one on top of the other, each about three
times too big for his face.

"Miss Rebecca!" he yelled as we climbed into the car. "Miss
Rebecca!"

I turned around. "Yes?"

"I love you."

"Thank you."

I turned back to Deborah, who was telling me how I shouldn't say
something or other around Zakariyya.

"Miss Rebecca! Miss Rebecca!" Alfred yelled again, slowly pushing both pairs of sunglasses down to the tip of his nose and wiggling
his eyebrows at me.

"You're mine," he said.

"Oh knock that off!" Deborah yelled, swatting at him from the
front seat. "Oh Lord, he just like his father, Mr. Ladies' Man." She
shook her head. "My son always out rippin and runnin them streets,
drinkin and druggin just like *his* father. I worry he gonna get himself
in trouble—I don't know what gonna happen to Little Alfred then.
I'm afraid he learnin too much already." Little Alfred was always
beating up on Davon, even though Davon was older and bigger, but
Davon never hit back without Deborah's permission.

When I asked the boys to tell me about their uncle Zakariyya,
Davon puffed up his chest, sucked in his nose so his nostrils vanished,
then yelled "GET THE HELL OUT OF HERE!" his voice deeper
than I thought possible for an eight-year-old. He and Alfred burst out
laughing and collapsed into a pile in the backseat. "Like one of them
wrestlers on TV!" Davon said, gasping for breath.

Alfred screamed and bounced in his seat. "WWF!! WWF!!"

Deborah looked at me and smiled. "Don't worry," she said. "I know how to handle him. I just keep remindin him to separate: Rebecca's not one of them researchers, she's not working for John Hopkin. She workin for herself. He kept sayin, 'I'm all right, I won't do nothing crazy.' But if I detect anything wrong we'll leave right back outta there."

We drove for a few blocks in silence, passing boarded-up storefronts, rows of fast-food restaurants and liquor stores. At one point, Davon pointed to his school and told us about the metal detectors and how they locked all the students inside during classes. Eventually Deborah leaned over to me and whispered, "Younger brother always felt like he was cheated out of life, because when my mother had him, four months later, that's when the sickness broke down on her. Brother's got a lot of anger. You just got to make sure you say his name right."

I'd been saying it wrong, she told me, and I couldn't do that in front of him. He pronounced it *Zuh-CAR-ee-uh*, not *Zack-a-RYE-uh*. Bobbette and Sonny had a hard time remembering that, so they called him Abdul, one of his middle names. But only when he wasn't around.

"Whatever you do, don't call him Joe," Deborah told me. "A friend of Lawrence's called him Joe one Thanksgiving and Zakariyya knocked that man out right into his mashed potatoes."

Zakariyya was about to turn fifty and lived in an assisted-living facility that Deborah had helped him get into when he was on the streets. He qualified because of his deafness and the fact that he was nearly blind without glasses. He hadn't lived there long, but was already on probationary status for being loud and aggressive with the other residents.

As Deborah and the boys and I walked from the car toward the front door, Deborah cleared her throat loudly and nodded toward a hulk of a man hobbling from the building in khaki pants. He was five feet eight inches tall and weighed just under four hundred pounds. He wore bright blue orthopedic sandals, a faded Bob Marley T-shirt, and a white baseball hat that said, HAM, BACON, SAUSAGE.

"Hey Zakariyya!" Deborah yelled, waving her hands above her head.

Zakariyya stopped walking and looked at us. His black hair was buzzed close to his head, his face smooth and youthful like Deborah's except for his brow, which was creased from decades of scowling. Beneath thick plastic glasses, his eyes were swollen, bloodshot, and surrounded by deep dark circles. One hand leaned on a metal cane identical to Deborah's, the other held a large paper plate with at least a pint of ice cream on it, probably more. Under his arm, he'd folded several newspaper ad sections.

"You told me you'd be here in an hour," he snapped.

"Uh . . . yeah . . . sorry," Deborah mumbled. "There wasn't any traffic."

"I'm not ready yet," he said, then grabbed the bundle of newspaper from under his arm and smacked Davon hard across the face with it. "Why'd you bring them?" he yelled. "You know I don't like no kids around."

Deborah grabbed Davon's head and pressed it to her side, rubbing his cheek and stammering that their parents had to work and no one else could take them, but she swore they'd be quiet, wouldn't they? Zakariyya turned and walked to a bench in front of his building without saying another word.

Deborah tapped me on the shoulder and pointed to another bench on the opposite side of the building's entrance, a good fifteen feet from Zakariyya. She whispered, "Sit over here with me," then yelled, "Come on boys, why don't you show Miss Rebecca how fast you can run!"

Alfred and Davon raced around the concrete cul-de-sac in front of Zakariyya's building, yelling, "Look at me! Look at me! Take my picture!"

Zakariyya sat eating his ice cream and reading his ads like we didn't exist. Deborah glanced at him every few seconds, then back to me, then the grandkids, then Zakariyya again. At one point she crossed her eyes and stuck her tongue out at Zakariyya, but he didn't see.

Finally, Zakariyya spoke.

"You got the magazine?" he asked, staring into the street.

Zakariyya had told Deborah he wanted to read the *Johns Hopkins Magazine* story I wrote about their mother before he'd talk to me, and he wanted me sitting next to him while he read it. Deborah nudged me toward his bench, then jumped up saying she and the boys would wait upstairs for us, because it was better if we talked outside in the nice weather rather than being cooped up alone inside. It was in the nineties with dizzying humidity, but neither of us wanted me going in that apartment alone with him.

"I'll be watching from that window up there," Deborah whispered. She pointed several floors up. "If anything funny starts, just wave and I'll come down."

As Deborah and the boys walked inside the building, I sat beside Zakariyya and started telling him why I was there. Without looking at me or saying a word, he took the magazine from my hand and began reading. My heart pounded each time he sighed, which was often.

"Damn!" he yelled suddenly, pointing at a photo caption that said Sonny was Henrietta's youngest son. "He ain't youngest! I am!" He slammed the magazine down and glared at it as I said of course I knew he was the youngest, and the magazine did the captions, not me.

"I think my birth was a miracle," he said. "I believe that my mother waited to go to the doctor till after I was born because she wanted to have me. A child born like that, to a mother full of tumors and sick as she was, and I ain't suffered no kinda physical harm from it? It's possible all this is God's handiwork."

He looked up at me for the first time since I'd arrived, then reached up and turned a knob on his hearing aid.

"I switched it off so I didn't have to listen to them fool children," he said, adjusting the volume until it stopped squealing. "I believe what them doctors did was wrong. They lied to us for twenty-five years, kept them cells from us, then they gonna say them things *donated* by our mother. Them cells was stolen! Those fools come take blood from us sayin they need to run tests and not tell us that all these years they done profitized off of her? That's like hanging a sign on our backs saying, 'I'm a sucker, kick me in my butt.' People don't know we just as po' as po'. They probably think by what our mother cells had did that we well off. I hope George Grey burn in hell. If he wasn't dead already, I'd take a black pitchfork and stick it up his ass."

Without thinking, almost as a nervous reflex I said, "It's George *Gey*, not Grey."

He snapped back, "Who cares what his name is? He always tellin people my mother name Helen Lane!" Zakariyya stood, towering over me, yelling, "What he did was wrong! Dead wrong. You leave that stuff up to God. People say maybe them takin her cells and makin them live forever to create medicines was what God wanted. But I don't think so. If He wants to provide a disease cure, He'd provide a cure of his own, it's not for man to tamper with. And you don't lie and clone people behind their backs. That's wrong—it's one of the most violating parts of this whole thing. It's like me walking in your bathroom while you in there with your pants down. It's the highest degree of disrespect. That's why I say I hope he burn in hell. If he were here right now, I'd kill him dead."

Suddenly, Deborah appeared beside me with a glass of water. "Just thought you might be thirsty," she said, her voice stern like *What the hell is going on here,* because she'd seen Zakariyya standing over me yelling.

"Everything okay out here?" she asked. "Y'all still reportin?"

"Yeah," Zakariyya said. But Deborah put her hand on his shoulder, saying maybe it was time we all went inside.

As we walked toward the front door of his building, Zakariyya turned to me. "Them doctors say her cells is so important and did all this and that to help people. But it didn't do no good for her, and it don't do no good for us. If me and my sister need something, we can't even go see a doctor cause we can't afford it. Only people that can get any good from my mother cells is the people that got money, and whoever sellin them cells—they get rich off our mother and we got nothing." He shook his head. "All those damn people didn't deserve her help as far as I'm concerned."

Zakariyya's apartment was a small studio with a sliver of a kitchen where Deborah and the boys had been watching us from a window. Zakariyya's belongings could have fit into the back of a pickup truck: a small Formica table, two wooden chairs, a full-sized mattress with no frame, a clear plastic bed skirt, and a set of navy sheets. No blankets, no pillows. Across from his bed sat a small television with a VCR balanced on top.

Zakariyya's walls were bare except for a row of photocopied pictures. The one of Henrietta with her hands on her hips hung next to the only other known picture of her: in it, she stands with Day in a studio sometime in the forties, their backs board-straight, eyes wide and staring ahead, mouths frozen in awkward non-smiles. Someone had retouched the photo and painted Henrietta's face an unnatural yellow. Beside it was a breathtaking picture of his sister Elsie, standing in front of a white porch railing next to a basket of dried flowers. She's about six years old, in a plaid jumper dress, white T-shirt, bobby socks and shoes, her hair loose from its braids, right hand gripping something against her chest. Her mouth hangs slightly open, brow creased and worried, both eyes looking to the far right of the frame, where Deborah imagines her mother was standing.

Zakariyya pointed to several diplomas hanging near the photos, for welding, refrigeration, diesel. "I got so many damn diplomas," he said, "but jobs pass me by because of my criminal record and everything, so I still got all kind of troubles." Zakariyya had been in and out of trouble with the law since he got out of jail, with various charges for assault and drunk and disorderly conduct.

"I think them cells is why I'm so mean," he said. "I had to start fightin before I was even a person. That's the only way I figure I kept them cancer cells from growin all over me while I was inside my mother. I started fightin when I was just a baby in her womb, and I never known nothin different."

Deborah thinks it was more than that. "That evil woman Ethel taught him hate," she said. "Beat every drop of it into his little body— put the hate of a murderer into him."

Zakariyya snorted when he heard Ethel's name. "Livin with that abusive crazy woman was worse than livin in prison!" he yelled, his eyes narrowing to slits. "It's hard to talk about what she did to me. When I get to thinkin about them stories, make me want to kill her, and my father. Cause of him I don't know where my mother buried. When that fool die, I don't wanna know where he buried neither. He need to get to a hospital? Let him catch a cab! Same with the rest of the so-called family who buried her. I don't never wanna see them niggers no more."

Deborah cringed. "See," she said, looking at me. "Everybody else never let him talk because he speak things the way he want to. I say let him talk, even if we be upset by what he's sayin. He's mad, gotta get it out, otherwise he gonna keep on keeping it, and it's gonna blow him right on up."

"I'm sorry," Zakariyya said. "Maybe her cells have done good for some people, but I woulda rather had my mother. If she hadn't been sacrificed, I mighta growed up to be a lot better person than I am now."

Deborah stood from the bed where she'd been sitting with her grandsons' heads on her lap. She walked over to Zakariyya and put her arm around his waist. "Come on walk us out to the car," she said. "I got something I want to give you."

Outside, Deborah threw open the back of her jeep and rummaged through blankets, clothes, and papers until she turned around holding the photo of Henrietta's chromosomes that Christoph Lengauer had given her. She smoothed her fingers across the glass, then handed it to Zakariyya.

"These supposed to be her cells?" he asked.

Deborah nodded. "See where it stained bright colors? That's where all her DNA at."

Zakariyya raised the picture to eye level and stared in silence. Deborah rubbed her hand on his back and whispered, "I think if anybody deserve that, it's you, Zakariyya."

Zakariyya turned the picture to see it from every angle. "You want me to have this?" he said finally.

"Yeah, like you to have that, put it on your wall," Deborah said.

Zakariyya's eyes filled with tears. For a moment the dark circles seemed to vanish, and his body relaxed.

"Yeah," he said, in a soft voice unlike anything we'd heard that day. He put his arm on Deborah's shoulder. "Hey, thanks."

Deborah wrapped her arms as far around his waist as she could reach, and squeezed. "The doctor who gave me that said he been working with our mother for his whole career and he never knew anything about where they came from. He said he was sorry."

Zakariyya looked at me. "What's his name?"

I told him, then said, "He wants to meet you and show you the cells."

Zakariyya nodded, his arm still around Deborah's shoulder. "Okay," he said. "That sounds good. Let's go for it." Then he walked slowly back to his building, holding the picture in front of him at eye level, seeing nothing ahead but the DNA in his mother's cells.

31

Hela, Goddess of Death

The day after I got home from our marathon visit, a man Deborah didn't know called her asking if she'd ride on a HeLa float in a black rodeo. He told her to be careful of people looking to find out where Henrietta's grave was because they might want to steal her bones, since her body was so valuable to science. Deborah told the man she'd been talking to me for a book, and he warned her not to talk to white people about her story. She panicked and called her brother Lawrence, who told her the man was right, so she left me a message saying she couldn't talk to me anymore. But by the time I got the message and called her back, she'd changed her mind.

"Everybody always yellin, 'Racism! Racism! That white man stole that black woman's cells! That white man killed that black woman!' That's crazy talk," she told me. "We all black and white and everything else—this isn't a race thing. There's two sides to the story, and that's what we want to bring out. Nothing about my mother is truth if it's about wantin to fry the researchers. It's not about punish the doctors or slander the hospital. I don't want that."

Deborah and I would go on like this for a full year. Each time I visited, we'd walk the Baltimore Harbor, ride boats, read science books together, and talk about her mother's cells. We took Davon and Alfred to the Maryland Science Center, where they saw a twenty-foot wall covered floor to ceiling with a picture of cells stained neon green and magnified under a microscope. Davon grabbed my hand and pulled me toward the wall of cells, yelling, "Miss Rebecca! Miss Rebecca! Is that Great-Grandma Henrietta?" People nearby stared as I said, "Actually, they might be," and Davon pranced around singing, "Grandma Henrietta famous! Grandma Henrietta famous!"

At one point, as Deborah and I walked along the cobblestone streets of Fell's Point late at night, she turned to me and without prompting said, "I'll bring them medical records out on my terms and when I think is right." She told me that the night she tackled her mother's medical records and ran home, she'd thought I was trying to steal them. She said, "I just need somebody I can trust, somebody that will talk to me and don't keep me in the dark." She asked me to promise I wouldn't hide anything from her. I promised I wouldn't.

Between trips, Deborah and I would spend hours each week talking over the phone. Occasionally someone would convince her she couldn't trust a white person to tell her mother's story, and she'd call me in a panic, demanding to know whether Hopkins was paying me to get information from her like people said. Other times she'd get suspicious about money, like when a genetics textbook publisher called offering her $300 for permission to print the photo of Henrietta. When Deborah said they had to give her $25,000 and they said no, she called me demanding to know who was paying me to write my book, and how much I was going to give her.

Each time I told her the same thing: I hadn't sold the book yet, so at that point I was paying for my research with student loans and credit cards. And regardless, I couldn't pay her for her story. Instead, I said, if the book ever got published, I would set up a scholarship fund for descendants of Henrietta Lacks. On Deborah's good days, she was excited about the idea. "Education is everything," she'd say. "If I'd had more of it, maybe this whole thing about my mother wouldn't have been so hard. That's why I'm always tellin Davon, 'Keep on studyin, learnin all you can.'" But on bad days, she'd think I was lying and cut me off again.

Those moments never lasted long, and they always ended with Deborah asking me to promise yet again that I'd never hide anything from her. Eventually I told her she could even come with me when I did some of my research if she wanted, and she said, "I want to go to centers and colleges and all that. Learning places. And I want to get the medical record and autopsy report on my sister."

I began sending her stacks of information I uncovered about her mother—scientific journal articles, photos of the cells, even an occasional novel, poem, or short story based on HeLa. In one, a mad scientist used HeLa as a biological weapon to spread rabies; another featured yellow house paint made of HeLa cells that could talk. I sent Deborah news of exhibits where several artists projected Henrietta's cells on walls, and one displayed a heart-shaped culture she'd grown by fusing her own cells with HeLa. With each packet, I sent notes explaining what each thing meant, clearly labeling what was fiction and what wasn't, and warning her about anything that might upset her.

Each time Deborah got a package, she'd call to talk about what she read, and gradually her panicked calls grew less frequent. Soon, after she realized I was the same age as her daughter, she started calling me "Boo," and insisted I buy a cell phone because she worried about me driving the interstates alone. Each time I talked to her brothers she'd yell at them, only half joking, saying, "Don't you try to take my reporter! Go get your own!"

When we met for our first trip, Deborah got out of her car wearing a black ankle-length skirt, black sandals with heels, and a black shirt covered with an open black cardigan. After we hugged, she said, "I got on my reporter clothes!" She pointed at my black button-up shirt, black pants, and black boots and said, "You always wear black, so I figured I should dress like you so I blend in."

For each trip, Deborah filled her jeep floor to ceiling with every kind of shoes and clothes she might need ("You never know when the weather gonna change"). She brought pillows and blankets in case we got stranded somewhere, an oscillating fan in case she got hot, plus all her haircutting and manicure equipment from beauty school, boxes of videotapes, music CDs, office supplies, and every document she had related to Henrietta. We always took two cars because Deborah didn't trust me enough to ride with me yet. I'd follow behind, watching her black driving cap bop up and down to her music. Sometimes, when we rounded curves or stopped at lights, I could hear her belting out, "Born to Be Wild," or her favorite William Bell song, "I Forgot to Be Your Lover."

Eventually, Deborah let me come to her house. It was dark, with thick closed curtains, black couches, dim lights, and deep brown wood-paneled walls lined with religious scenes on blacklight posters. We spent all our time in her office, where she slept most nights instead of the bedroom she shared with Pullum—they fought a lot, she told me, and needed some peace.

Her room was about six feet wide, with a twin bed against one wall and a small desk directly across from it, nearly touching the bed. On top of the desk, stacked beneath reams of paper, boxes of envelopes, letters, and bills was her mother's Bible, its pages warped, cracking with age, and spotted with mold, her mother's and sister's hair still tucked inside.

Deborah's walls were covered floor to ceiling with colorful photos of bears, horses, dogs, and cats she'd torn from calendars, as well as nearly a dozen bright felt squares she and Davon had made by hand. One was yellow with THANK YOU JESUS FOR LOVING ME written in big letters; another said PROPHECIES FULFILLED and was covered with coins made of tinfoil. A shelf at the head of her bed was crammed with videotapes of infomercials: for a Jacuzzi, an RV, a trip to Disneyland. Nearly every night Deborah would say, "Hey Davon, you want to go on vacation?" When he nodded yes she'd ask, "Where you want to go, Disneyland, spa, or RV trip?" They'd watched each tape many times.

At the end of one visit, I showed Deborah how to get online with an old computer someone had given her years earlier, then taught her to use Google. Soon she started taking Ambien—a prescription sleep aid—and sitting up nights in a drugged haze, listening to William Bell on headphones, Googling "Henrietta" and "HeLa."

Davon referred to Deborah's Ambien as "dummy medicine," because it made her wander the house in the middle of the night like a zombie, talking nonsense and trying to cook breakfast by chopping cereal with a butcher knife. When he stayed with her, Davon often woke up in the middle of the night to find Deborah sleeping at her computer, head down and hands on the keyboard. He'd just push her off the chair into bed and tuck her in. When Davon wasn't there, Deborah often woke up with her face on the desk, surrounded by a mountain of pages that spilled from her printer onto the floor: scientific articles, patent applications, random newspaper articles and blog posts, including many that had no connection to her mother but used the words *Henrietta* or *lacks* or *Hela*.

And, surprisingly, there were many of the latter. Hela is the native name for the country of Sri Lanka, where activists carry signs demanding "Justice for the Hela Nation." It's the name of a defunct German tractor company and an award-winning shih-tzu dog; it's a seaside resort in Poland, an advertising firm in Switzerland, a Danish boat where people gather to drink vodka and watch films, and a Marvel comic book character who appears in several online games: a seven-foot-tall, half-black, half-white goddess who's part dead and part alive, with "immeasurable" intelligence, "superhuman" strength, "godlike" stamina and durability, and five hundred pounds of solid muscle. She's responsible for plagues, sickness, and catastrophes; she's immune to fire, radiation, toxins, corrosives, disease, and aging. She can also levitate and control people's minds.

When Deborah found pages describing Hela the Marvel character, she thought they were describing her mother, since each of Hela's traits in some way matched what Deborah had heard about her mother's cells. But it turned out the sci-fi Hela was inspired by the ancient Norse goddess of death, who lives trapped in a land between hell and the living. Deborah figured that goddess was based on her mother too.

One day, around three o'clock in the morning, my phone rang as I slept, feverish with flu. Deborah yelled on the other end, "I told you London cloned my mother!" Her voice was slow and slurred from Ambien.

She'd Googled *HeLa, clone, London,* and *DNA,* and gotten thousands of hits with summaries like this, from an online chat-room discussion about HeLa cells: "Each contains a genetic blueprint for constructing Henrietta Lacks. . . . Can we clone her?" Her mother's name showed up under headlines like CLONING and HUMAN FARMING, and she thought those thousands of hits were proof that scientists had cloned thousands of Henriettas.

"They didn't clone her," I said. "They just made copies of her cells. I promise."

"Thanks Boo, I'm sorry I woke you," she cooed. "But if they cloning her cells, does that mean someday they could clone my mother?"

"No," I said. "Good night."

After several weeks of finding Deborah unconscious, with her phone in her hand, or face on the keyboard, Davon told his mother he needed to stay at his grandmother's house all the time, to take care of her after she took her medicine.

Deborah took an average of fourteen pills a day, which cost her about $150 each month after her husband's insurance, plus Medicaid and Medicare. "I think it's eleven prescriptions," she told me once, "maybe twelve. I can't keep track, they change all the time." One for acid reflux went from $8 one month to $135 the next, so she stopped taking it, and at one point her husband's insurance canceled her prescription coverage, so she started cutting her pills in half to make them last. When the Ambien ran out, she stopped sleeping until she got more.

She told me her doctors started prescribing the drugs in 1997 after what she referred to as "the Gold Digger Situation," which she refused to tell me about. That was when she'd applied for Social Security disability, she said, which she only got after several court appearances.

"Social Security people said everything was all in my head," she told me. "They ended up sending me to about five psychiatrist and a bunch of doctors. They say I'm paranoia, I'm schizophrenia, I'm nervous. I got anxiety, depression, degenerating kneecaps, bursitis, bulged discs in my back, diabetes, osteoporosis, high blood pressure, cholesterol. I don't know all of what's wrong with me by name," she said. "I don't know if anyone do. All I know is, when I get in that mood and I get frightened, I hide."

That's what happened the first time I called, she said. "I was all excited, sayin I want a book written about my mother. Then things just started going in my head and I got scared.

"I know my life could be better and I wish it was," she told me. "When people hear about my mother cells they always say, 'Oh y'all could be rich! Y'all gotta sue John Hopkin, y'all gotta do this and that.' But I don't *want* that." She laughed. "Truth be told, I can't get mad at science, because it help people live, and I'd be a mess without it. I'm a walking drugstore! I can't say nuthin bad about science, but I won't lie, I *would* like some health insurance so I don't got to pay all that money every month for drugs my mother cells probably helped make."

Eventually, as Deborah grew comfortable with the Internet, she started using it for more than terrifying herself in the middle of the night. She made lists of questions for me and printed articles about research done on people without their knowledge or consent—from a vaccine trial in Uganda to the testing of drugs on U.S. troops. She started organizing information into carefully labeled folders: one about cells, another about cancer, another full of definitions of legal terms like *statute of limitations* and *patient confidentiality.* At one point she stumbled on an article called "What's Left of Henrietta Lacks?" that infuriated her by saying Henrietta had probably gotten HPV because she "slept around."

"Them people don't know nothing about science," she told me. "Just havin HPV don't mean my mother was loose. Most people got it—I read about it on the Internet."

Then, in April 2001, nearly a year after we first met, Deborah called to tell me that "the president of a cancer club" had called wanting to put her on stage at an event honoring her mother. She was worried, she said, and she wanted me to find out if he was legit.

He turned out to be Franklin Salisbury Jr., president of the National Foundation for Cancer Research. He'd decided to hold the foundation's 2001 conference in Henrietta's honor. On September 13, seventy top cancer researchers from around the world would gather to present their research, he said, and hundreds of people would attend, including the mayor of Washington, D.C., and the surgeon general. He hoped Deborah would speak there, and accept a plaque in her mother's honor.

"I understand that the family feels very abused," he told me. "We can't give them money, but I'm hoping this conference will set the historic record straight and help make them feel better, even if we are fifty years late."

When I explained this to Deborah, she was ecstatic. It would be just like Pattillo's conference in Atlanta, she said, only bigger. She immediately started planning what she'd wear and asking questions about what the researchers would be talking about. And she worried again about whether she'd be safe on stage, or whether there'd be a sniper waiting for her.

"What if they think I'm going to cause trouble about them taking the cells or something?"

"I don't think you need to worry about that," I said. "The scientists are excited to meet you." Besides, I told her, it was going to be in a federal building with high security.

"Okay," she said. "But first I want to go see my mother cells, so I know what everybody's talkin about at the conference."

When we hung up I went to call Christoph Lengauer, the cancer researcher who'd given Deborah the painted chromosome picture, but before I could dig out his number, my phone rang again. It was Deborah, crying. I thought she was panicking, changing her mind about seeing the cells. But instead she wailed, "Oh my baby! Lord help him, they got him with fingerprints on a pizza box."

Her son Alfred and a friend had been on a crime spree, robbing at least five liquor stores at gunpoint. Security cameras caught Alfred on tape yelling at a store clerk and waving a bottle of Wild Irish Rose above his head. He'd stolen a twelve-ounce bottle of beer, one bottle of Wild Irish Rose, two packs of Newport cigarettes, and about a hundred dollars in cash. The police arrested him in front of his house and threw him in the car while his son, Little Alfred, watched from the lawn.

"I still want to go see them cells," Deborah said, sobbing. "I ain't gonna let this stop me from learning about my mother and my sister."

32

"All That's My Mother"

By the time Deborah was ready to see her mother's cells for the first time, Day couldn't come. He'd said many times that he wanted to see his wife's cells before he died, but he was eighty-five, in and out of the hospital with heart and blood pressure problems, and he'd just lost a leg to diabetes. Sonny had to work, and Lawrence said he wanted to talk to a lawyer about suing Hopkins instead of seeing the cells, which he referred to as "a multibillion-dollar corporation."

So on May 11, 2001, Deborah, Zakariyya, and I agreed to meet at the Hopkins Jesus statue to go see Henrietta's cells. Earlier that morning, Deborah had warned me that Lawrence was convinced Hopkins was paying me to gather information about the family. He'd already called her several times that day saying he was coming to get the materials she'd collected related to her mother. So Deborah locked them in her office, took the key with her, and called me saying, "Don't tell him where you are or go see him without me."

When I arrived at the Jesus, it stood just as it had when Henrietta visited it some fifty years earlier, looming more than ten feet tall beneath a tiered dome, pupil-less marble eyes staring straight ahead, arms outstretched and draped in stone robes. At Jesus's feet, people had thrown piles of change, wilted daisies, and two roses—one fresh with thorns, the other cloth with plastic dewdrops. His body was gray-brown and dingy, except for his right foot, which glowed a polished white from decades of hands rubbing it for luck.

Deborah and Zakariyya weren't there, so I leaned against a far wall, watching a doctor in green scrubs kneel before the statue and pray as others brushed its toe on their way into the hospital without looking or breaking stride. Several people stopped to write prayers in oversized books resting on wooden pedestals near the statue: "Dear Heavenly Father: If it is your will let me speak to Eddie this one last time." "Please help my sons conquer their addictions." "I ask you to provide my husband and I with jobs." "Lord thank you for giving me another chance."

I walked to the statue, my heels echoing on marble, and rested my hand on its big toe—the closest I'd ever come to praying. Suddenly Deborah was beside me, whispering, "I hope He's got our back on this one." Her voice was utterly calm, her usual nervous laugh gone.

I told her I did too.

Deborah closed her eyes and began to pray. Then Zakariyya appeared behind us and let out a deep laugh.

"He can't do nothin to help you now!" Zakariyya yelled. He'd gained weight since I'd seen him last, and his heavy gray wool pants and thick blue down coat made him look even bigger. The black plastic arms of his glasses were so tight they'd etched deep grooves into his head, but he couldn't afford new ones.

He looked at me and said, "That sister of mine, she crazy for not wantin money from them cells."

Deborah rolled her eyes and hit his leg with her cane. "Be good or you can't come see the cells," she said.

Zakariyya stopped laughing and followed as we headed toward Christoph Lengauer's lab. Minutes later, Christoph walked toward us through the lobby of his building, smiling, hand outstretched. He was in his mid-thirties, with perfectly worn denim jeans, a blue plaid shirt, and shaggy light brown hair. He shook my hand and Deborah's, then reached for Zakariyya's. But Zakariyya didn't move.

"Okay!" Christoph said, looking at Deborah. "It must be pretty hard for you to come into a lab at Hopkins after what you've been through. I'm really glad to see you here." He spoke with an Austrian accent, which made Deborah wiggle her eyebrows at me when he turned to press the elevator call button. "I thought we'd start in the freezer room so I can show you how we store your mother's cells, then we can go look at them alive under a microscope."

"That's wonderful," Deborah said, as though he'd just said something entirely ordinary. Inside the elevator, she pressed against Zakariyya, one hand leaning on her cane, the other gripping her tattered dictionary. When the doors opened, we followed Christoph single file through a long narrow hall, its walls and ceiling vibrating with a deep whirring sound that grew louder as we walked. "That's the ventilation system," Christoph yelled. "It sucks all the chemicals and cells outside so we don't have to breathe them in."

He threw open the door to his lab with a sweeping *ta-da* motion and waved us inside. "This is where we keep all the cells," he yelled over a deafening mechanical hum that made Deborah's and Zakariyya's hearing aids squeal. Zakariyya's hand shot up and tore his from his ear. Deborah adjusted the volume on hers, then walked past Christoph into a room filled wall-to-wall with white freezers stacked one on top of the other, rumbling like a sea of washing machines in an industrial laundromat. She shot me a wide-eyed, terrified look.

Christoph pulled the handle of a white floor-to-ceiling freezer, and it opened with a hiss, releasing a cloud of steam into the room. Deborah screamed and jumped behind Zakariyya, who stood expressionless, hands in his pockets.

"Don't worry," Christoph yelled, "it's not dangerous, it's just cold. They're not minus twenty Celsius like your freezers at home, they're minus eighty. That's why when I open them smoke comes out." He motioned for Deborah to come closer.

"It's all full of her cells," he said.

Deborah loosened her grip on Zakariyya and inched forward until the icy breeze hit her face, and she stood staring at thousands of inch-tall plastic vials filled with red liquid.

"Oh God," she gasped. "I can't believe all that's my mother." Zakariyya just stared in silence.

Christoph reached into the freezer, took out a vial, and pointed to the letters *H-e-L-a* written on its side. "There are millions and millions of her cells in there," he said. "Maybe billions. You can keep them here forever. Fifty years, a hundred years, even more—then you just thaw them out and they grow."

He rocked the vial of HeLa cells back and forth in his hand as he started talking about how careful you have to be when you handle them. "We have an extra room just for the cells," he said. "That's important. Because if you contaminate them with anything, you can't really use them anymore. And you don't want HeLa cells to contaminate other cultures in a lab."

"That's what happened over in Russia, right?" Deborah said.

He did a double take and grinned. "Yes," he said. "Exactly. It's great you know about that." He explained how the HeLa contamination problem happened, then said, "Her cells caused millions of dollars in damage. Seems like a bit of poetic justice, doesn't it?"

"My mother was just getting back at scientists for keepin all them secrets from the family," Deborah said. "You don't mess with Henrietta—she'll sic HeLa on your ass!"

Everyone laughed.

Christoph reached into the freezer behind him, grabbed another vial of HeLa cells, and held it out to Deborah, his eyes soft. She stood stunned for a moment, staring into his outstretched hand, then grabbed the vial and began rubbing it fast between her palms, like she was warming herself in winter.

"She's cold," Deborah said, cupping her hands and blowing onto the vial. Christoph motioned for us to follow him to the incubator where he warmed the cells, but Deborah didn't move. As Zakariyya and Christoph walked away, she raised the vial and touched it to her lips.

"You're famous," she whispered. "Just nobody knows it."

Christoph led us into a small laboratory crammed full of microscopes, pipettes, and containers with words like BIOHAZARD and DNA written on their sides. Pointing to the ventilation hoods covering his tables, he said, "We don't want cancer all over the place, so this sucks all the air to a filtration system that catches and kills any cells that are floating around."

He explained what culture medium was, and how he moved cells from freezer to incubator to grow. "Eventually they fill those huge bottles in the back," he said, pointing to rows of gallon-sized jugs. "Then we do our experiments on them, like we find a new drug for cancer, pour it onto the cells, and see what happens." Zakariyya and Deborah nodded as he told them how drugs go through testing in cells, then animals, and finally humans.

Christoph knelt in front of an incubator, reached inside, and pulled out a dish with HeLa growing in it. "They're really, really small, the cells," he said. "That's why we go to the microscope now so I can show them to you." He flipped power switches, slid the dish onto the microscope's platform, and pointed to a small monitor attached to the microscope. It lit up a fluorescent green, and Deborah gasped.

"It's a pretty color!"

Christoph bent over the microscope to bring the cells into focus, and an image appeared on the screen that looked more like hazy green pond water than cells.

"At this magnification you can't see much," Christoph said. "The screen is just boring because the cells are so small, even with a microscope you can't see them sometimes." He clicked a knob and zoomed in to higher and higher magnifications until the hazy sea of green turned into a screen filled with hundreds of individual cells, their centers dark and bulging.

"Oooo," Deborah whispered. "There they are." She reached out and touched the screen, rubbing her finger from one cell to the next.

Christoph traced the outline of a cell with his finger. "All this is one cell," he said. "It kinda looks like a triangle with a circle in the middle, you see that?"

He grabbed a piece of scrap paper and spent nearly a half-hour drawing diagrams and explaining the basic biology of cells as Deborah asked questions. Zakariyya turned up his hearing aid and leaned close to Christoph and the paper.

"Everybody always talking about cells and DNA," Deborah said at one point, "but I don't understand what's DNA and what's her cells."

"Ah!" Christoph said, excited, "DNA is what's *inside* the cell! Inside each nucleus, if we could zoom in closer, you'd see a piece of DNA that looked like this." He drew a long, squiggly line. "There's forty-six of those pieces of DNA in every human nucleus. We call those chromosomes—those are the things that were colored bright in that big picture I gave you."

"Oh! My brother got that picture hanging on his wall at home next to our mother and sister," Deborah said, then looked at Zakariyya. "Did you know this is the man who gave you that picture?"

Zakariyya looked to the ground and nodded, the corners of his mouth turning up into a barely perceptible smile.

"Within the DNA in that picture is all the genetic information that made Henrietta *Henrietta*," Christoph told them. "Was your mother tall or short?"

"Short."

"And she had dark hair, right?"

We all nodded.

"Well, all that information came from her DNA," he said. "So did her cancer—it came from a DNA mistake."

Deborah's face fell. She'd heard many times that she'd inherited some of the DNA inside those cells from her mother. She didn't want to hear that her mother's cancer was in that DNA too.

"Those mistakes can happen when you get exposed to chemicals or radiation," Christoph said. "But in your mother's case, the mistake was caused by HPV, the genital warts virus. The good news for you is that children don't inherit those kinds of changes in DNA from their parents—they just come from being exposed to the virus."

"So we don't have the thing that made her cells grow forever?" Deborah asked. Christoph shook his head. "Now you tell me after all these years!" Deborah yelled. "Thank God, cause I *was* wonderin!"

She pointed at a cell on the screen that looked longer than the others. "This one is cancer, right? And the rest are her normal ones?"

"Actually, HeLa is *all* just cancer," Christoph said.

"Wait a minute," she said, "you mean none of our mother *regular* cells still livin? Just her cancer cells?"

"That's right."

"Oh! See, and all this time I thought my mother *regular* cells still livin!"

Christoph leaned over the microscope again and began moving the cells quickly around the screen until he shrieked, "Look, there! See that cell?" He pointed to the center of the monitor. "See how it has a big nucleus that looks like it's almost pinched in half in the middle? That cell is dividing into two cells right before our eyes! And both of those cells will have your mother's DNA in them."

"Lord have mercy," Deborah whispered, covering her mouth with her hand.

Christoph kept talking about cell division, but Deborah wasn't listening. She stood mesmerized, watching one of her mother's cells divide in two, just as they'd done when Henrietta was an embryo in her mother's womb.

Deborah and Zakariyya stared at the screen like they'd gone into a trance, mouths open, cheeks sagging. It was the closest they'd come to seeing their mother alive since they were babies.

After a long silence, Zakariyya spoke.

"If those our mother's cells," he said, "how come they ain't black even though she was black?"

"Under the microscope, cells don't have a color," Christoph told him. "They all look the same—they're just clear until we put color on them with a dye. You can't tell what color a person is from their cells." He motioned for Zakariyya to come closer. "Would you like to look at them through the microscope? They look better there."

Christoph taught Deborah and Zakariyya how to use the microscope, saying, "Look through like this . . . take your glasses off . . . now turn this knob to focus." Finally the cells popped into view for Deborah. And through that microscope, for that moment, all she could see was an ocean of her mother's cells, stained an ethereal fluorescent green.

"They're *beautiful*," she whispered, then went back to staring at the slide in silence. Eventually, without looking away from the cells, she said, "God, I never thought I'd see my mother under a microscope— I never dreamed this day would ever come."

"Yeah, Hopkins pretty much screwed up, I think," Christoph said.

Deborah bolted upright and looked at him, stunned to hear a scientist—one at Hopkins, no less—saying such a thing. Then she looked back into the microscope and said, "John Hopkin is a school for learning, and that's important. But this *is* my mother. Nobody seem to get that."

"It's true," Christoph said. "Whenever we read books about science, it's always HeLa *this* and HeLa *that*. Some people know those are the initials of a person, but they don't know who that person is. That's important history."

Deborah looked like she wanted to hug him. "This is amazing," she said, shaking her head and looking at him like he was a mirage.

Suddenly, Zakariyya started yelling something about George Gey. Deborah thumped her cane on his toe and he stopped in midsentence.

"Zakariyya has a lot of anger with all this that's been goin on," she told Christoph. "I been trying to keep him calm. Sometime he explode, but he's trying."

"I don't blame you for being angry," Christoph said. Then he showed them the catalog he used to order HeLa cells. There was a long list of the different HeLa clones anyone could buy for $167 a vial.

"You should get that," Christoph said to Deborah and Zakariyya.

"Yeah, right," Deborah said. "What I'm gonna do with a vial of my mother cells?" She laughed.

"No, I mean you should get the money. At least some of it."

"Oh," she said, stunned. "That's okay. You know, when people hear about who HeLa was, first thing they say is, 'Y'all should be millionaires!' "

Christoph nodded. "Her cells are how it all started," he said. "Once there is a cure for cancer, it's definitely largely because of your mother's cells."

"Amen," Deborah said. Then, without a hint of anger, she told him, "People always gonna be makin money from them cells, nothing we can do about that. But we not gonna get any of it."

Christoph said he thought that was wrong. Why not treat valuable cells like oil, he said. When you find oil on somebody's property, it doesn't automatically belong to them, but they do get a portion of the profits. "No one knows how to deal with this when it comes to cells today," he said. "When your mother got sick, doctors just did what they wanted and patients didn't ask. But nowadays patients want to know what's going on."

"Amen," Deborah said again.

Christoph gave them his cell phone number and said they could call any time they had questions about their mother's cells. As we walked toward the elevator, Zakariyya reached up and touched Christoph on the back and said thank you. Outside, he did the same to me, then turned to catch the bus home.

Deborah and I stood in silence, watching him walk away. Then she put her arm around me and said, "Girl, you just witnessed a miracle."

33

The Hospital for the Negro Insane

There were several things I'd promised Deborah we'd do together: seeing her mother's cells was first; figuring out what happened to Elsie was second. So the day after we visited Christoph's lab, Deborah and I set out on a weeklong trip that would start at Crownsville, where we hoped to find her sister's medical records, then go through Clover and end in Roanoke, at the house where Henrietta was born.

It was Mother's Day, which had always been a sad day for Deborah, and this one hadn't started well. She'd planned to take her grandson Alfred to see his father in jail before we left town. But her son had called saying he didn't want Deborah or Little Alfred visiting until he could see them without looking through glass. He told her he wanted to learn about his grandmother, Henrietta, and asked Deborah to send him whatever information we found on our trip.

"I been waiting for him to say that his whole life," she told me, crying. "I just didn't want him to have to get locked up in prison to do it." But once again, she said, "I'm not gonna let that stop me.

I just want to focus on the good, like seein my mother cells, and learnin about my sister." So we drove to Crownsville in our separate cars.

I don't know what I expected the former Hospital for the Negro Insane to look like, but it certainly wasn't what we found. Crownsville Hospital Center was on a sprawling 1,200-acre campus, with bright green hills, perfectly mowed lawns, walking paths, weeping cherry trees, and picnic tables. Its main building was red brick with white columns, its porch decorated with wide chairs and chandeliers. It looked like a nice place to sip mint juleps or sweet tea. One of the old hospital buildings was now a food bank; others housed the Police Criminal Investigation Division, an alternative high school, and a Rotary club.

Inside the main building, we walked past empty offices in a long, empty white hallway, saying, "Hello?" and "Where is everybody?" and "This place is weird." Then, at the end of the hall was a white door covered with years' worth of dirt and handprints. It had the words MEDICAL RECORDS stenciled across it in broken block letters. Beneath that, in smaller letters, it said NO THOROUGHFARE.

Deborah gripped the door handle and took a deep breath. "We ready for this?" she asked. I nodded. She grabbed my arm with one hand, threw the door open with the other, and we stepped inside.

We found ourselves in a thick white metal cage that opened into the Medical Records room—an empty, warehouse-sized room with no staff, no patients, no chairs, no visitors, and no medical records. Its windows were bolted shut and covered with wire and dirt, its gray carpet bunched in ripples from decades of foot traffic. A waist-high cinder-block wall ran the length of the room, separating the waiting area from the area marked AUTHORIZED PERSONNEL ONLY, where several rows of tall metal shelves stood empty.

"I can't believe this," Deborah whispered. "All them records is gone?" She ran her hand along the empty shelves, mumbling, "Nineteen fifty-five was the year where they killed her. . . . I want them records. . . . I know it wasn't good. . . . Why else would they get rid of them?"

No one had to tell us something awful had happened at Crownsville—we could feel it in the walls.

"Let's go find someone who can tell us something," I said.

We wandered into another long hallway, and Deborah began screaming. "Excuse me! We need to find the medical record! Does anyone know where it is?"

Eventually a young woman poked her head out of an office and pointed us down the hall to another office, where someone pointed us to yet another. Finally we found ourselves in the office of a tall man with a thick white Santa Claus beard and wild, bushy eyebrows. Deborah charged over to him, saying, "Hi, I'm Deborah, and this is my reporter. You may have heard of us, my mama's in history with the cells, and we need to find some medical record."

The man smiled. "Who was your mother," he asked, "and what are the cells?"

We explained why we were there, and he told us that the current medical records were in another building, and that there wasn't much history left at Crownsville. "I wish we had an archivist," he said. "I'm afraid I'm as close as you'll get."

His name was Paul Lurz, and he was the hospital's director of performance and improvement, but he also happened to be a social worker who'd majored in history, which was his passion. He motioned for us to come sit in his office.

"There wasn't much funding for treating blacks in the forties and fifties," he said. "I'm afraid Crownsville wasn't a very nice place to be back then." He looked at Deborah. "Your sister was here?"

She nodded.

"Tell me about her."

"My father always say she never went past a child in her head," she said, reaching into her purse for a crumpled copy of Elsie's death certificate, which she began reading slowly out loud. "Elsie Lacks . . . cause of death (a) respiratory failure (b) epilepsy (c) cerebral palsy. . . . Spent five years in Crownsville State Hospital." She handed Lurz the picture of her sister that Zakariyya had hanging on his wall. "I don't believe my sister had all that."

Lurz shook his head. "She doesn't look like she has palsy in this picture. What a lovely child."

"She did have them seizures," Deborah said. "And she couldn't never learn how to use the toilet. But I think she was just deaf. Me and all my brothers got a touch of nerve deafness on account of our mother and father being cousins and having the syphilis. Sometimes I wonder, if somebody taught her sign language, maybe she'd still be alive."

Lurz sat in his chair, legs crossed, looking at the photo of Elsie. "You have to be prepared," he told Deborah, his voice gentle. "Sometimes learning can be just as painful as not knowing."

"I'm ready," Deborah said, nodding.

"We had a serious asbestos problem," he said. "Most of our records from the fifties and earlier were contaminated. Instead of cleaning each page of the records to save them, the administration decided to have them carted away in bags and buried."

He walked to a storage closet near his desk, its walls lined with shelves and file cabinets. In the back corner he'd crammed a small desk, facing the wall. Lurz had been working at Crownsville since 1964, when he was a student intern in his twenties, and he had a habit of collecting potentially historic documents: patient records, copies of old admissions reports that caught his attention—an infant admitted blind in one eye with facial deformities and no family, a child institutionalized without any apparent psychiatric disorder.

Lurz disappeared into the closet and began muttering amid loud clunking and shuffling noises. "There were a few . . . I just had them out a couple weeks ago . . . Ah! Here we go." He walked out of the closet carrying a stack of oversized books with thick leather spines and dark green cloth covers. They were warped with age, coated in dust, and filled with thick, yellowed paper.

"These are autopsy reports," he said, opening the first book as the scent of mildew filled the room. He'd found them while rummaging in the basement of an abandoned building at the hospital sometime in the eighties, he said. When he'd first opened them, hundreds of bugs scurried from the pages onto his desk.

Between 1910, when the hospital opened, and the late fifties, when the records were found to be contaminated, tens of thousands of patients passed through Crownsville. Their records—if they'd survived—could have filled Lurz's small storage room several times over. Now this stack was all that was left at Crownsville.

Lurz pulled out a volume that included some reports from 1955, the year Elsie died, and Deborah squealed with excitement.

"What did you say her full name was?" Lurz asked, running his finger down a list of names written in careful script next to page numbers.

"Elsie Lacks," I said, scanning the names over his shoulder as my heart raced. Then, in a daze, I pointed to the words *Elsie Lacks* on the page and said, "Oh my God! There she is!"

Deborah gasped, her face suddenly ashen. She closed her eyes, grabbed my arm to steady herself, and started whispering, "Thank you Lord . . . Thank you Lord."

"Wow. This really surprises me," Lurz said. "It was very unlikely she'd be in here."

Deborah and I began hopping around and clapping. No matter what the record said, at least it would tell us something about Elsie's life, which we figured was better than knowing nothing at all.

Lurz opened to Elsie's page, then quickly closed his eyes and pressed the book to his chest before we could see anything. "I've never seen a picture in one of these reports," he whispered.

He lowered the book so we all could see, and suddenly time seemed to stop. The three of us stood, our heads nearly touching over the page, as Deborah cried, "Oh my baby! She look just like my daughter! . . . She look just like Davon! . . . She look just like my father! . . . She got that smooth olive Lacks skin."

Lurz and I just stared, speechless.

In the photo, Elsie stands in front of a wall painted with numbers for measuring height. Her hair, which Henrietta once spent hours combing and braiding, is frizzy, with thick mats that stop just below the five-foot mark behind her. Her once-beautiful eyes bulge from her head, slightly bruised and almost swollen shut. She stares somewhere just below the camera, crying, her face misshapen and barely recognizable, her nostrils inflamed and ringed with mucus; her lips—swollen to nearly twice their normal size—are surrounded by a deep, dark ring of chapped skin; her tongue is thick and protrudes from her mouth. She appears to be screaming. Her head is twisted unnaturally to the left, chin raised and held in place by a large pair of white hands.

"She doesn't want her head like that," Deborah whispered. "Why are they holding her head like that?"

No one spoke. We all just stood there, staring at those big white hands wrapped around Elsie's neck. They were well manicured and feminine, pinky slightly raised—hands you'd see in a commercial for nail polish, not wrapped around the throat of a crying child.

Deborah laid her old picture of Elsie as a young girl next to the new photo.

"Oh, she was beautiful," Lurz whispered.

Deborah ran her finger across Elsie's face in the Crownsville photo. "She looks like she wonderin where I'm at," she said. "She look like she needs her sister."

The photo was attached to the top corner of Elsie's autopsy report, which Lurz and I began reading, saying occasional phrases out loud: "diagnosis of idiocy" . . . "directly connected with syphilis" . . . "self-induced vomiting by thrusting fingers down her throat for six months prior to death." In the end, it said, she was "vomiting coffee-ground material," which was probably clotted blood.

Just as Lurz read the phrase "vomiting coffee-ground material" out loud, a short, round, balding man in a dark business suit stormed into the room telling me to stop taking notes and demanding to know what we were doing there.

"This is the family of a patient," Lurz snapped. "They're here to look at the patient's medical records."

The man paused, looking at Deborah, then at me: a short black woman in her fifties, and a taller white woman in her twenties. Deborah gripped her cane and stared him in the eye with a look that just begged him to mess with her. She reached into her bag and pulled out three pieces of paper: her birth certificate, Elsie's birth certificate, and the legal document giving her power of attorney over Elsie, something she'd spent months getting, just in case anyone tried to stop her from doing precisely what we were doing.

She handed them to the man, who grabbed the autopsy report book and started reading. Deborah and I glared at him, both so furious at him for trying to stop us that neither of us realized he was one of the only hospital officials who'd ever tried to protect the Lacks family's privacy.

"Can Deborah get a copy of that autopsy report?" I asked Lurz.

"Yes, she can," he said, "if she submits a written request." He grabbed a piece of paper from his desk and handed it to Deborah.

"What am I supposed to write?" she asked.

Lurz began reciting: "I, Deborah Lacks . . ."

Within moments she had an official medical record request on a torn piece of paper. She handed it to Lurz and told him, "I need a good blowed-up copy of that picture, too."

Before Lurz left to make photocopies, with the bald man close behind, he handed me a stack of photos and documents to look at while he was gone. The first document in the stack was a *Washington Post* article from 1958, three years after Elsie's death, with the headline:

OVERCROWDED HOSPITAL "LOSES" CURABLE PATIENTS
Lack of Staff at Crownsville Pushes Them to Chronic Stage

The second I read the title, I flipped the article facedown in my lap. For a moment I considered not showing it to Deborah. I thought maybe I should read it first, so I could prepare her for whatever awful thing we were about to learn. But she grabbed it from my hand and read the headline out loud, then looked up, her eyes dazed.

"This is nice," she said, pointing to a large illustration that showed a group of men in various states of despair, holding their heads, lying on the floor, or huddling in corners. "I'd like to have this for my wall." She handed it back to me and asked me to read it out loud.

"Are you sure?" I asked. "This is probably going to say some pretty upsetting things. Do you want me to read it first and tell you what it says?"

"No," she snapped. "Like he told us, they didn't have the money to take care of black people." She walked behind me to follow along over my shoulder as I read, then she scanned the page and pointed to several words on the page: "Gruesome?" she said. "Fearsome black wards?"

The Crownsville that Elsie died in was far worse than anything Deborah had imagined. Patients arrived from a nearby institution packed in a train car. In 1955, the year Elsie died, the population of Crownsville was at a record high of more than 2,700 patients, nearly eight hundred above maximum capacity. In 1948, the only year figures were available, Crownsville averaged one doctor for every 225 patients, and its death rate was far higher than its discharge rate. Patients were locked in poorly ventilated cell blocks with drains on the floors instead of toilets. Black men, women, and children suffering with everything from dementia and tuberculosis to "nervousness," "lack of self-confidence," and epilepsy were packed into every conceivable space, including windowless basement rooms and barred-in porches. When they had beds, they usually slept two or more on a twin mattress, lying head to foot, forced to crawl across a sea of sleeping bodies to reach their beds. Inmates weren't separated by age or sex, and often included sex offenders. There were riots and homemade weapons. Unruly patients were tied to their beds or secluded in locked rooms.

I later learned that while Elsie was at Crownsville, scientists often conducted research on patients there without consent, including one study titled "Pneumoencephalographic and skull X-ray studies in 100 epileptics." Pneumoencephalography was a technique developed in 1919 for taking images of the brain, which floats in a sea of fluid. That fluid protects the brain from damage, but makes it very difficult to X-ray, since images taken through fluid are cloudy. Pneumoencephalography involved drilling holes into the skulls of research subjects, draining the fluid surrounding their brains, and pumping air or helium into the skull in place of the fluid to allow crisp X-rays of the brain through the skull. The side effects—crippling headaches, dizziness, seizures, vomiting—lasted until the body naturally refilled the skull with spinal fluid, which usually took two to three months. Because pneumoencephalography could cause permanent brain damage and paralysis, it was abandoned in the 1970s.

There is no evidence that the scientists who did research on patients at Crownsville got consent from either the patients or their parents. Based on the number of patients listed in the pneumoencephalography study and the years it was conducted, Lurz told me later, it most likely involved every epileptic child in the hospital, including Elsie. The same is likely true of at least one other study, called "The Use of Deep Temporal Leads in the Study of Psychomotor Epilepsy," which involved inserting metal probes into patients' brains.

Soon after Elsie's death, a new warden took over at Crownsville and began releasing hundreds of patients who'd been institutionalized unnecessarily. The *Washington Post* article quoted him saying, "The worst thing you can do to a sick person is close the door and forget about him."

When I read that line out loud, Deborah whispered, "We didn't forget about her. My mother died . . . nobody told me she was here. I would have got her out."

As we left Crownsville, Deborah thanked Lurz for the information, saying, "I've been waiting for this a long, long time, Doc." When he asked if she was okay, her eyes welled with tears and she said, "Like I'm always telling my brothers, if you gonna go into history, you can't do it with a hate attitude. You got to remember, times was different."

When we got outside, I asked Deborah if she was sure she was all right. She just laughed like I was crazy. "It was such a good idea we decided to stop here," she said, then hurried to the parking lot, climbed into her car, and rolled the window down. "Where we goin next?"

Lurz had mentioned that any other remaining old records from Crownsville were stored at the Maryland State Archives in Annapolis, about seven miles away. He didn't think they'd have any from the fifties, but figured it wouldn't hurt to look.

"We goin to Annapolis see if they got more of my sister medical records?"

"I don't know if that's a good idea," I said. "Don't you want a break?"

"No way!" she yelled. "We got lots more reportin to do—we just gettin hot now!" She screeched off in her car, smiling and waving the new picture of her sister out the window at me as I jumped in my car to follow.

About ten minutes later, as we pulled into the parking lot of the State Archives, Deborah bounced in the seat of her car, gospel music blaring so loud I could hear it with my windows up. When we walked inside, she went straight to the reception desk, reached into her bag, pulled out her mother's medical records, and waved them in the air above her head, saying, "They call my mother HeLa! She's in all the computers!"

I was relieved when the receptionist said the archives didn't have Elsie's medical records. I didn't know how much more Deborah could take, and I was scared of what we'd find.

The rest of the day was a blur. As we drove to Clover, each time we stopped, Deborah leapt from her car, clutching the new photo of her sister and thrusting it into the face of every person we met: a woman on a street corner, the man pumping our gas, a pastor at a small church, our waitresses. Each time, she said, "Hi, my name's Deborah and this is my reporter, you probably heard of us, my mama's in history with the cells, and we just found this picture of my sister!"

Each time, the reaction was the same: sheer horror. But Deborah didn't notice. She just smiled and laughed, saying, "I'm so happy our reportin is going so good!"

As the day went on, the story behind the picture grew more elaborate. "She's a little puffy from cryin because she misses my mother," she said at one point. Another time she told a woman, "My sister's upset because she's been looking for me but can't find me."

Occasionally she'd pull over to the side of the road and motion for me to pull up beside her so she could tell me various ideas she'd come up with as she drove. At one point she'd decided she needed to get a safe deposit box for her mother's Bible and hair; later she asked if she needed to copyright Henrietta's signature so no one would steal it. At a gas station, while we waited in line for the bathroom, she pulled a hammer from her backpack and said, "I wish the family would give me the home-house so I can make it a historical place. But they won't, so I'm gonna take the doorknob so at least I have something from it."

At one point, Deborah climbed from her car looking near tears. "I been havin a hard time keepin my eyes on that road," she said. "I just keep lookin at the picture of my sister." She'd been driving with both of Elsie's pictures on the passenger seat beside her, staring at them as she drove. "I can't get all these thoughts outta my head. I just keep thinkin about what she must've gone through in those years before she died."

I wanted to take the picture from her so she'd stop torturing herself with it, but she wouldn't have let me if I'd tried. Instead, I just kept saying maybe we should go home, it had been an intense couple days, and perhaps she wasn't ready for so much reporting at once. But each time, Deborah told me I was crazy if I thought she was stopping now. So we kept going.

At several points during the day, Deborah said I should take her mother's medical records into my hotel room when we stopped for the night. "I know you'll have to look at every page, take notes and everything, cause you need all the facts." And finally, when we checked into a hotel somewhere between Annapolis and Clover around nine o'clock at night, she gave them to me.

"I'm going to sleep," she said, walking into the room next to mine. "Knock yourself out."

34

The Medical Records

A few minutes later, Deborah pounded on my door. She'd changed into an enormous white T-shirt that hung past her knees—on it was a picture of a stick-figure woman taking cookies out of an oven, and the word GRANDMA in big childlike print.

"I decided I'm not going to bed," she said matter-of-factly. "I want to look at that stuff with you." She was jittery and twitchy, like she'd just had several shots of espresso. In one hand she clutched the Crownsville picture of Elsie; with the other she grabbed the bag filled with her mother's medical records off the dresser where I'd put it. She dumped the bag's contents on my bed just as she'd done the first night we met.

"Let's get busy," she said.

There were more than a hundred pages, many of them crumpled, folded, or torn, all of them out of order. I stood staring for a long moment, stunned and overwhelmed, then said maybe we could sort through it together, then I could find somewhere to photocopy what I'd need.

"No!" Deborah yelled, then smiled a nervous smile. "We can just read it all here and you can take notes."

"That would take days," I said.

"No it won't," Deborah said, climbing on all fours across the pile of papers, and sitting cross-legged in the center of the bed.

I pulled up an armchair, opened my laptop, and started sorting. There was a land deed from the small chunk of Clover property Deborah bought with two thousand dollars from her father's asbestos settlement. There was a 1997 newspaper mug shot of Lawrence's son with a caption that said, WANTED. LAWRENCE LACKS, ROBBERY W/DEADLY WEAPON. There were order forms for buying HeLa cells online, receipts, newsletters from Deborah's church, and seemingly endless copies of the photo of Henrietta, hands on hips. And there were dozens of notebook pages where Deborah had written definitions of scientific and legal terms, and poems about her life:

> *cancer*
> *check up*
> *can't afford*
> *white and rich get it*
> *my mother was black*
> *black poor people don't have the money to*
>   *pay for it*
> *mad yes I am mad*
> *we were used by taking our blood and lied to*
> *We had to pay for our own medical, can you*
>   *relieve that.*
> *John Hopkin Hospital and all other places,*
>   *that has my mother cells, don't give her*
> *Nothing.*

As I read, Deborah grabbed several photocopied pages from a genealogy how-to book and held them up for me to see, saying, "That's how I knew to get power of attorney and bring all that stuff to get my sister information at Crownsville. They didn't know who they was foolin' with!" As she talked, she watched my hands moving through the pile of papers.

I held a page of the records close to my face to make out the small script, then began reading out loud, " 'This twenty-eight-year-old' . . . something . . . I can't read the handwriting . . . 'positive Rh.' " The entry was dated November 2, 1949.

"Oh wow!" I said suddenly. "This is three days before you were born—your mom's pregnant with you here."

"What? Oh my god!" Deborah screamed, snatching the paper and staring at it, mouth wide. "What else does it say?"

It was a normal checkup, I told her. "Look here," I said, pointing at the page. "Her cervix is two centimeters dilated . . . She's getting ready to have you."

Deborah bounced on the bed, clapped her hands, and grabbed another page from the medical records.

"Read this one!"

The date was February 6, 1951. "This is about a week after she first went to the hospital with her cervical cancer," I said. "She's waking up from anesthesia after getting her biopsy. It says she feels fine."

For the next few hours, Deborah pulled papers off the pile for me to read and sort. One moment she'd screech with joy over a fact I'd found, the next she'd panic over a new fact that didn't sit well, or at the sight of me holding a page of her mother's medical records. Each time she panicked, she'd pat the bed and say, "Where's my sister autopsy report?" or "Oh no, where'd I put my room key?"

Occasionally she stashed papers under the pillow, then pulled them out when she decided it was okay for me to see them. "Here's my mother autopsy," she said at one point. A few minutes later she handed me a page she said was her favorite because it had her mother's signature on it—the only piece of Henrietta's handwriting on record. It was the consent form she'd signed before her radium treatment, when the original HeLa sample was taken.

Eventually, Deborah grew quiet. She lay on her side and curled herself around the Crownsville picture of Elsie for so long, I thought she'd fallen asleep. Then she whispered, "Oh my god. I don't like the way she got her neck." She held up the picture and pointed to the white hands.

"No," I said. "I don't like that either."

"I know you was hopin I didn't notice that, weren't you?"

"No. I knew you noticed."

She laid her head back down again. We kept on like this for hours, me reading and taking notes, Deborah staring at Elsie's picture in long silences broken only by her sparse commentary: "My sister look scared." . . . "I don't like that look on her face." . . . "She was chokin herself?" . . . "I guess after she realized she wasn't going to see my mother no more, she just gave up." Occasionally she shook her head hard, like she was trying to snap herself out of something.

Eventually I leaned back in my chair and rubbed my eyes. It was the middle of the night and I still had a big pile of paper to sort through.

"You might think about getting yourself another copy of your mother's medical record and stapling it with all the pages in order to keep it all straight," I said.

Deborah squinted at me, suddenly suspicious. She moved across the room to the other bed, where she lay on her stomach and started reading her sister's autopsy report. A few minutes later, she jumped up and grabbed her dictionary.

"They diagnosed my sister with idiocy?" she said, then started reading the definition out loud. " 'Idiocy: utterly senseless or foolish.' " She threw down the dictionary. "That's what they say was wrong with my sister? She had *foolish*? She was an idiot? How can they do that?"

I told her that doctors used to use the word *idiocy* to refer to mental retardation, and to the brain damage that accompanied hereditary syphilis. "It was sort of a generic word to describe someone who was slow," I said.

She sat down next to me and pointed to a different word in her sister's autopsy report. "What does this word mean?" she asked, and I told her. Then her face fell, her jaw slack, and she whispered, "I don't want you puttin that word in the book."

"I won't," I said, and then I made a mistake. I smiled. Not because I thought it was funny, but because I thought it was sweet that she was protective of her sister. She'd never told me something was off limits for the book, and this was a word I would never have included—to me, it didn't seem relevant. So I smiled.

Deborah glared at me. "Don't you put that in the book!" she snapped.

"I *won't*," I told her, and I meant it. But I was still smiling, now more from nervousness than anything else.

"You're lying," Deborah yelled, flipping off my tape recorder and clenching her fists.

"I'm not, I swear, look, I'll say it on tape and you can sue me if I use it." I clicked the recorder on, said into the mic that I wouldn't put that word in the book, then turned it off.

"You're lying!" she yelled again. She jumped off the bed and stood over me, pointing a finger in my face. "If you're not lying, why did you smile?"

She started frantically stuffing papers into her canvas bags as I tried to explain myself and talk her down. Suddenly she threw the bag on the bed and rushed toward me. Her hand hit my chest hard as she slammed me against the wall, knocking me breathless, my head smacking the plaster.

"Who you working for?" she snapped. "John Hopkin?"

"What? No!" I yelled, gasping for breath. "You know I work for myself."

"Who sent you? Who's paying you?" she yelled, her hand still holding me against the wall. "Who paid for this room?"

"We've been through this!" I said. "Remember? Credit cards? Student loans?"

Then, for the first time since we met, I lost my patience with Deborah. I jerked free of her grip and told her to get the fuck off me and chill the fuck out. She stood inches from me, staring wild-eyed again for what felt like minutes. Then, suddenly, she grinned and reached up to smooth my hair, saying, "I never seen you mad before. I was starting to wonder if you was even human cause you never cuss in front of me."

Then, perhaps as an explanation for what just happened, she finally told me about Cofield.

"He was a good pretender," she said. "I told him I would walk through fire alive before I would let him take my mother medical records. I don't want nobody else to have them. Everybody in the world got her cells, only thing we got of our mother is just them records and her Bible. That's why I get so upset about Cofield. He was trying to take one of the only things I really got from my mother."

She pointed at my laptop on the bed and said, "I don't want you typin every word of it into your computer either. You type what you need for the book, but not everything. I want people in our family to be the only ones who have all them records."

After I promised I wouldn't copy all the records, Deborah said she was going to bed again, but for the next several hours, she knocked on my door every fifteen or twenty minutes. The first time she reeked of peaches and said, "I just had to go to my car for my lotion so I thought I'd say hi." Each time it was something else: "I forgot my nail file in the car!" . . . "*X-Files* is on!" . . . "I'm suddenly thinking about pancakes!" Each time she knocked, I opened my door wide so she could see the room and the medical records looking just as they had when she left.

The last time she knocked, she stormed past me into the bathroom and leaned over the sink, her face close to the mirror. "Am I broken out?" she yelled. I walked into the bathroom, where she stood pointing to a quarter-sized welt on her forehead. It looked like a hive.

She turned and pulled her shirt down so I could see her neck and back, which were covered in red welts.

"I'll put some cream on it," she said. "I should probably take my sleeping pill." She went back to her room and a moment later the volume on her TV went up. Screaming and crying and gunfire poured out of the television all night, but I didn't see her again until six o'clock in the morning—one hour after I'd gone to sleep—when she knocked on my door yelling, "Free continental breakfast!"

My eyes were red and swollen with dark circles under them, and I was still wearing my clothes from the day before. Deborah looked at me and laughed.

"We're a mess!" she said, pointing to the hives now covering her face. "Lord, I was so anxious last night. I couldn't do anything with myself so I painted my fingernails." She held out her hands for me to see. "I did a *horrible* job!" she said, laughing. "I think I did it after I took my pill."

Her nails and much of the skin around them were bright fire-engine red. "From a distance it looks okay," she said. "But I'd get fired if I was still doin nails for a living."

We walked down to the lobby for our free breakfast. As Deborah wrapped a handful of mini-muffins in a napkin for later, she looked up at me and said, "We're okay, Boo."

I nodded and said I knew. But at that point I wasn't sure of anything.

35

Soul Cleansing

By later that day, the hives had spread across Deborah's back, her cheeks were splotchy and red, and long welts filled the spaces beneath each eye. Both lids were swollen and shining like she'd covered them in blood-red shadow. I asked again and again if she was okay and said maybe we should stop somewhere so she could see a doctor. But she just laughed.

"This happens all the time," she said. "I'm fine. I just need some Benadryl." She bought a bottle that she kept in her purse and swigged from all day. By noon, about a third of it was gone.

When we got to Clover, we walked along the river, down Main Street, and through Henrietta's tobacco field. And we visited the home-house, where Deborah said, "I want you to take a picture of me here with my sister."

She stood in front of the house, turned both photos of Elsie so they faced me, and held them to her chest. She had me take pictures of her and Elsie on the stump of what used to be Henrietta's favorite oak tree and in front of Henrietta's mother's tombstone. Then she knelt on the ground, next to the sunken strips of earth where she imagined her mother and sister were buried. "Take one of me and my sister by her and my mother grave," she said. "It'll be the only picture in the world with the three of us almost together."

Finally we ended up at Henrietta's sister Gladys's house, a small yellow cabin with rocking chairs on its porch. Inside we found Gladys sitting in her dark wood-paneled living room. It was warm out, sweatshirt weather, but Gladys had her double-wide black wood-stove burning so hot, she sat beside it wiping sweat from her forehead with tissue. Her hands and feet were gnarled from arthritis, her back so bent her chest nearly touched her knees unless she propped herself up with an elbow. She wore no underwear, only a thin nightgown that had ridden above her waist from hours in her wheelchair.

She tried to straighten her gown to cover herself when we walked in, but her hands couldn't grasp it. Deborah pulled it down for her, saying, "Where everybody at?"

Gladys said nothing. In the next room, her husband moaned from a hospital bed, just days from death.

"Oh right," Deborah said, "they at work ain't they?"

Gladys said nothing, so Deborah raised her voice loud to make sure Gladys could hear: "I got a Internet!" she yelled. "I'm going to get a web page up about my mother and hopefully be getting some donations and funding so I can come back down here put a monument up on her grave and turn that old home-house into a museum that will remind people of my mother down here!"

"What you put in there?" Gladys asked, like Deborah was crazy.

"Cells," Deborah said. "Cells so people can see her multiply."

She thought for a moment. "And a great big picture of her, and maybe one of them wax statues. Plus some of them old clothes and that shoe in the house. All that stuff mean a whole lot."

Suddenly the front door opened and Gladys's son Gary came inside yelling, "Hey Cuz!" Gary was fifty, with that smooth Lacks skin, a thin mustache and soul patch, and a gap between his front teeth that the girls loved. He wore a red and blue short-sleeved rugby shirt that matched his blue and red jeans and sneakers.

Deborah squealed, threw her arms around Gary's neck, and pulled the photo of Elsie from her pocket. "Look what we got from Crownsville! It's my sister!" Gary stopped smiling and reached for the picture.

"That's a bad shot," Deborah said. "She's crying cause it's cold."

"How about showing him that picture of her on the porch when she was a kid?" I said. "That's a good one." Gary looked at me like, *What the hell is going on here?*

"That picture's got her a little upset," I said.

"I understand why," he whispered.

"Plus she just saw her mother's cells for the first time," I told him.

Gary nodded. Over the years, he and I had spent many hours talking; he understood Deborah and what she'd been through more than anyone else in her family.

Deborah pointed to the hives on her face. "I'm having a reaction, swellin up and breakin out. I'm crying and happy at the same time." She started pacing back and forth, her face shining with sweat as the woodstove clanged and seemed to suck most of the oxygen from the room. "All this stuff I'm learning," she said, "it make me realize that I *did* have a mother, and all the tragedy she went through. It hurts but I wanna know more, just like I wanna know about my sister. It make me feel closer to them, but I do miss them. I wish they were here."

Keeping his eyes on Deborah, Gary walked across the room, sat in an oversized recliner, and motioned for us to join him. But Deborah didn't sit. She paced back and forth across the linoleum floor, picking the red polish off her nails and talking an incoherent stream about a murder she'd heard about on the news and the traffic in Atlanta. Gary's eyes followed her from one side of the room to the other, intense and unblinking.

"Cuz," he said finally. "Please sit."

Deborah raced over to a rocking chair not far from Gary, threw herself into it, and started rocking violently, thrusting her upper body back and forth and kicking her feet like she was trying to flip the chair over.

"You wouldn't believe what we been learning!" she said. "They injected my mother's cells with all kinds of, uh, poisons and stuff to test if they'd kill people."

"Dale," Gary said, "do something for yourself."

"Yeah, I'm tryin," she said. "You know they shot her cells into murderers in prison?"

"I mean to relax," Gary said. "Do something to *relax* yourself."

"I can't help it," Deborah said, waving him off with her hand. "I worry all the time."

"Like the Bible said," Gary whispered, "man brought nothing into this world and he'll carry nothing out. Sometime we care about stuff too much. We worry when there's nothing to worry about."

In a moment of clarity, Deborah nodded, saying, "And we bring our own body down by doing it."

"You don't seem so good right now, Cuz. Make some time for yourself," Gary said. "When I get in my car and drive, don't have to be going nowhere, circles is fine by me. Just got to have time to relax with the road under me. Everybody needs something like that."

"If I ever get any money," Deborah said, "I'll get an RV where I can go back and forth and I don't have to be in the same place ever. Can't nobody bother you when you're movin."

She stood up and started pacing again.

"Only time I really relax is when I'm drivin down here," she said. "But this time I just be drivin along the whole time thinking about what happened to my sister and my mother."

The moment Deborah said the words *sister* and *mother,* her face got redder and she started to panic. "You know they shot my mother cells into space and blew her up with nuclear bombs? They even did that thing . . . what do you call it . . . um . . . *cloning!* . . . that's right, they did that cloning on her."

Gary and I shot each other a nervous glance and both started talking at once, scrambling to bring her back from wherever she was going.

"There are no clones," I said. "Remember?"

"You don't have to be fearful," Gary said. "The word of God said if we honor our father and mother, we can live long upon the earth, and you doing that, you honoring your mother." He smiled and closed his eyes. "I love this scripture that's in Psalms," he told her. "It says even if our father and our mother fall sick, the Lord take care of you. Even if you lose everybody like your mother and your sister, God's love will never turn His back on you."

But Deborah didn't hear any of it.

"You wouldn't believe it," she said. "You know they mixed her with mice to make a human-mouse? They say she's not even human anymore!" She laughed a loud, manic laugh and ran to the window. "Holy cuss!" she yelled, "is it raining out there?"

"Much needed rain," Gary whispered, rocking back and forth.

Deborah grabbed the blue ribbon keychain that always hung around her neck. It said WWJD. "What is this," she said, "a radio station? I never heard of WWJD." She started yanking it off her neck.

"Come on, Cuz, it means 'What Would Jesus Do,'" Gary said. "You know that."

Deborah stopped fussing with the keys and collapsed back into the chair. "Can you believe they even gave her that AID virus and injected her into monkeys?" She stared at the floor, rocking violently, her chest rising and falling fast with each breath.

Gary sat, calmly rocking in his chair, watching Deborah's every move, like a doctor studying a patient. "Don't make yourself sick over something you can't do nothin about," Gary whispered to Deborah as she rubbed the welts on her eyes. "It's not worth it . . . you got to let the Lord handle it." His eyes drooped closed as he mumbled, "What is Deborah doing for Deborah?"

When she didn't answer he looked at me and said, "I was talking to God just now—he's trying to make me say stuff, trying to make me move." Deborah called Gary The Disciple because he had a habit of channeling the Lord in the middle of a conversation. It started about twenty years earlier, when he was thirty—one minute he was busy with booze and women, the next he'd had several heart attacks and bypasses, and he woke up preaching.

"I been tryin to keep Him out of this because we've got company," he said, flashing me a bashful grin. "But sometimes He just won't let me keep Him out."

Gary's brown eyes went vacant, unfocused, as he stood slowly from his chair, spread his arms wide, and reached toward Deborah, who struggled to her feet, hobbled toward him, and wrapped her arms around his waist. The moment she touched him, his upper body seized like he'd been electrocuted. His arms thrust closed, hands clasping each side of Deborah's head, palms to her jaw, fingers spread from the back of her skull to the bridge of her nose. Then he started shaking. He squeezed Deborah's face to his chest as her shoulders heaved in silent sobs, and tears rolled from Gary's eyes.

As they rocked back and forth, Gary tipped his head to the sky, and began singing in a hauntingly beautiful baritone.

"Welcome, into this place. . . . Welcome, into this broken vessel." His singing, quiet at first, grew louder with each word until it filled the house and poured into the tobacco fields. *"You desire to abide in the praises of your people, so I lift my hand, and I lift my heart, and I offer up this praise unto ya, Lord."*

"You're welcome into this broken vessel, Lord," he whispered, squeezing Deborah's head in his palms. His eyes shot open and closed, and he began to preach, sweat pouring from his face.

"That you said in your word Lord, that the BELIEVER would lay hands on the sick, and that they shall RECOVER!" His voice rose and fell, from a whisper to a yell and back. "I REALIZE God that TONIGHT there's just *some things* doctors CANNOT DO!"

"Amen Lord," Deborah mumbled, face pressed to his chest, voice muffled.

"We thank ya tonight," Gary whispered. "Because we need your help with them CELLS, Lord . . . we need your help liftin the BURDEN of them cells from this woman! Lift this burden, Lord, take it away, we don't NEED it!"

Deborah started convulsing in Gary's arms, weeping and whispering, "Thank ya, Lord . . . Thank ya, Lord." Gary squeezed his eyes tight, and yelled along with her, "THANK YOU, LORD! THANK YOU FOR TONIGHT!" Their voices grew louder together, until Gary stopped, tears and sweat pouring from his face onto Deborah as she screamed, "Thank you Jesus!" and let loose with a chorus of hallelujahs and praise Gods. Gary swayed back and forth, breaking into song again, his voice deep and old, as if coming from the generations who worked his tobacco fields before him: *"I know the Lord been good, yoooooooooooh . . . I know the Lord been good."*

"Real good," Deborah whispered.

"He's put food on my table . . ." Gary dropped his voice, humming as Deborah spoke: "Show me which way to go, Lord," she said. "Show me where you want me to go with these cells, Lord, *please.* I'll do anything you want me to do, Lord, just help me with this BURDEN. I can't do it alone—I thought I could. But I can't TAKE it, Lord."

Mmmmmmm mmmmmmm mmmmmmm, Gary hummed.

"Thank you Lord for giving me this information about my mother and my sister, but please HELP ME, cause I know I can't handle this burden by myself. Take them CELLS from me, Lord, take that BURDEN. Get it off and LEAVE it there! I can't carry it no more, Lord. You wanted me to give it to you and I just didn't want to, but you can have it now, Lord. You can HAVE IT! Hallelujah, amen."

For the first time since Gary stood from his chair, he looked straight at me.

I'd been watching all this from a recliner a few feet away, dumbfounded, terrified to move or make noise, frantically scribbling notes. In any other circumstance I might have thought the whole thing was crazy. But what was happening between Gary and Deborah at that moment was the furthest thing from crazy I'd seen all day. As I watched, all I could think was, *Oh my god . . . I did this to her.*

Gary stared into my eyes as he hugged Deborah's sobbing body and whispered to her, "You're not alone."

Looking at me, Gary said, "She can't handle the burden of these cells no more, Lord! She can't do it!" Then he raised his arms above Deborah's head and yelled, "LORD, I KNOW you sent Miss Rebecca to help LIFT THE BURDEN of them CELLS!" He thrust his arms toward me, hands pointed at either side of my head. "GIVE THEM TO HER!" he yelled. "LET HER CARRY THEM."

I sat frozen, staring at Gary, thinking, *Wait a minute, that wasn't supposed to happen!*

Deborah stepped away from Gary's embrace, shaking her head, wiping her eyes, and yelling, "Phew!" They both laughed. "Thanks, Cuz," she said, "I feel so light!"

"Some things you got to release," Gary said. "The more you hold them in, the worse you get. When you release them, they got to go somewhere else. The Bible says He can carry all that burden."

She reached up and touched his face. "You always know what I need. You know how to take care of me."

"It's not so much that I see it, but He sees it," Gary said, smiling. "I didn't know all that was coming out my mouth. That was the Lord talking to you."

"Well, hallelujah," Deborah said, giggling. "I'm comin back tomorrow for some more of this! Amen!"

It had been drizzling outside for hours, but suddenly rain pounded the tin roof and turned to hail so loud that it sounded like applause. The three of us walked to the front door to look.

"It's the Lord saying he heard us," Gary said, smiling. "He got the faucet turned on high to clean you out, Cuz!"

"Praise the Lord!" Deborah yelled.

Gary hugged Deborah good-bye, then hugged me. Deborah grabbed her long black raincoat, opened it wide, and raised it above her like an umbrella, nodding for me to come under with her. She let the coat fall onto both of our heads, then put her arm tight around my shoulders.

"You ready for some soul cleansing?" she yelled, opening the door.

36

Heavenly Bodies

The next morning Deborah's hives had gone down some, but her eyes were still swollen, so she decided she needed to go home to see her doctor. I stayed behind in Clover because I wanted to talk to Gary about the night before. When I walked into his living room he was standing on a plastic folding chair in a bright turquoise shirt, changing a lightbulb.

"I can't get that beautiful song out of my head," I told him. "I've been singing it all morning." Then I hummed a few bars: *Welcome into this place . . . welcome into this broken vessel.*

Gary jumped off the chair, laughing and raising his eyebrows at me.

"Now why do you think that's stuck in your head?" he asked. "I know you don't like to think about it, but that's the Lord telling you something."

He said it was a hymn, then ran from the living room and came back carrying a soft blue Bible with large gold lettering across its front. "I want you to have this," he told me, tapping the cover with his finger. "He died for us that we might have the right to eternal life.

A lot of people don't believe that. But you *can* have eternal life. Just look at Henrietta."

"You believe Henrietta is in those cells?"

He smiled and looked down his nose at me like, *silly child.* "Those cells *are* Henrietta," he said, taking back the Bible and opening it to the book of John. "Read that," he said, pointing to a chunk of text. I started reading to myself and he covered the Bible with his hand. "Out loud," he said.

So I read aloud from the Bible, for the first time in my life: "Those who believe in me will live, even though they die; and those who live and believe in me will never die."

Gary flipped to another passage for me to read: "Someone will ask, 'How can the dead be raised to life? What kind of body will they have?' You fool! When you plant a seed in the ground, it does not sprout to life unless it dies. And what you plant is a bare seed . . . not the full-bodied plant that will later grow up. God provides that seed with the body he wishes; he gives each seed its own proper body."

"Henrietta was chosen," Gary whispered. "And when the Lord chooses an angel to do his work, you never know what they going to come back looking like."

Gary pointed at another passage and told me to keep reading. "There are heavenly bodies and earthly bodies, the beauty that be-longs to heavenly bodies is different from the beauty that belongs to earthly bodies."

When Christoph projected Henrietta's cells on the monitor in his lab a few days earlier, Deborah said, "They're beautiful." She was right. Beautiful and otherworldly—glowing green and moving like water, calm and ethereal, looking precisely like heavenly bodies might look. They could even float through the air.

I kept reading: "This is how it will be when the dead are raised to life. When the body is buried, it is mortal; when raised, it will be immortal. There is, of course, a physical body, so there has to be a spiritual body."

"HeLa?" I asked Gary. "You're saying HeLa is her spiritual body?"

Gary smiled and nodded.

In that moment, reading those passages, I understood completely how some of the Lackses could believe, without doubt, that Henrietta had been chosen by the Lord to become an immortal being. If you believe the Bible is the literal truth, the immortality of Henrietta's cells makes perfect sense. *Of course* they were growing and surviving decades after her death, *of course* they floated through the air, and *of course* they'd led to cures for diseases and been launched into space. Angels are like that. The Bible tells us so.

For Deborah and her family—and surely many others in the world—that answer was so much more concrete than the explanation offered by science: that the immortality of Henrietta's cells had something to do with her telomeres and how HPV interacted with her DNA. The idea that God chose Henrietta as an angel who would be reborn as immortal cells made a lot more sense to them than the explanation Deborah had read years earlier in Victor McKusick's genetics book, with its clinical talk of HeLa's "atypical histology" and "unusually malignant behavior." It used phrases like "the tumor's singularity" and called the cells "a reservoir of morphologic, biochemical, and other information."

Jesus told his followers, "I give them eternal life, and they shall never die." Plain, simple, to the point.

"You better be careful," Gary told me. "Pretty soon you're gonna find yourself converted."

"I doubt it," I told him, and we both laughed.

He slid the Bible from my hands and flipped to another passage, then handed it back, pointing at one sentence: "Why do you who are here find it impossible to believe that God raises the dead?"

"You catch my drift?" he said, smiling a mischievous grin.

I nodded, and Gary closed the Bible in my hands.

37

"Nothing to Be Scared About"

When Deborah got to her doctor's office, her blood pressure and blood sugar were so high, her doctor was amazed she hadn't had a stroke or heart attack while we were in Clover. With levels like hers, he said, she could still have one any minute. Suddenly her strange behavior on the trip seemed less strange. Confusion, panic, and incoherent speech are all symptoms of extremely high blood pressure and blood sugar, which can lead to heart attack and stroke. So is redness and swelling, which could explain why her red welts didn't go away despite all the Benadryl she drank.

The doctor told her she needed to avoid stress completely, so we decided she should stop coming on research trips with me. But she insisted I call her from the road to tell her what she was missing. For the next several months, as I continued my research, I told Deborah only the good things I found: stories about Henrietta dancing and watching the boys play baseball at Cliff's house, details about her family history from county records and wills.

But we both knew the break from HeLa wouldn't last—Deborah was still scheduled to give a talk at the National Foundation for

Cancer Research conference in honor of Henrietta. She was determined to do it, even though she was terrified by the idea of getting up on stage, so she started spending her days planning her speech.

One afternoon, in the midst of preparing for the conference, she called me to say she'd decided she wanted to go to school. "I keep thinkin, maybe if I understood some science, then the story about my mother and sister wouldn't scare me so much," she said. "So I'm just gonna do it." Within days, she'd called several local community centers and found one that offered adult education classes, and signed up to take math and reading placement tests.

"Once I get tenth-grade level, I'm ready to go on to college!" she told me. "Can you imagine? Then I can understand all that science about my mother!" She thought about becoming a dental assistant, but was leaning toward radiation technologist so she could study cancer and help patients who were getting radiation treatment like her mother.

As the conference approached, Deborah was calm, but I wasn't. I kept asking, "Are you sure you want to do this?" and "How's your blood pressure?" and "Does your doctor know you're doing this?" She kept telling me she was fine, that even her doctor said so.

Deborah took her placement tests for school and registered for the classes she'd need to get herself up to tenth-grade level and qualify for the community college classes she wanted to take. She called me, giddy, screaming, "I start a week from today!"

But everything else seemed to be spiraling in the wrong direction. A few days before the conference, Lawrence and Zakariyya called yelling again about how she shouldn't talk to anyone, and saying they wanted to sue every scientist who'd ever worked on Henrietta's cells. Sonny told them to stay out of it, saying, "All she doin now is goin places to speak and learn—y'all don't want to do that, so just leave her alone." But Lawrence insisted Deborah give him the records she'd gathered on their mother.

Then her son Alfred called from prison, saying he'd finally be going on trial right after the conference, and the charges now included armed robbery and attempted murder. That same day, Deborah got a call about one of Lawrence's sons who'd been arrested for robbery and was in the same jail as Alfred.

"The Devil's been busy, girl," she told me. "I love them boys, but I'm not gonna let nobody upset me right now."

The next morning was September 11, 2001.

I called Deborah around eight in the morning, saying I was leaving my home in Pittsburgh, and headed to the conference in Washington, D.C. Less than an hour later, the first plane hit the World Trade Center. A reporter friend called my cell phone and told me the news, saying, "Don't go to D.C., it's not safe." I turned my car around as the second plane hit, and by the time I got home, the TV was filled with footage of the Pentagon's wreckage and buildings throughout D.C. being evacuated, including the Ronald Reagan Building, where the conference reception to honor Henrietta was supposed to be held.

I called Deborah, and she answered in a panic. "It's just like Pearl Harbor all over again," she said. "And Oklahoma! There's no way I'm going to D.C. now." But there was no need. With airlines and Washington shut down, the NFCR canceled the Henrietta Lacks conference, with no plan to reschedule.

For the next several days, Deborah and I talked many times as we both struggled to make sense of the attacks, and Deborah tried to accept the idea that the conference had been canceled. She was depressed, and worried that it would take another ten years for someone to honor her mother.

Then, on Sunday morning, five days after September 11, Deborah went to church to pray for Alfred, whose trial was only a few days away, and to ask that the Henrietta Lacks conference be rescheduled. She sat in the front pew in a red dress suit, hands folded in her lap, listening to her husband preach about September 11. About an hour into the service, Deborah realized she couldn't move her arm.

Davon, who was now nine years old, always sat in the choir and watched his grandmother during church. For a moment, as Deborah's face began to sag and her body slumped, Davon thought maybe she'd accidentally taken her sleeping pill before coming to church. Deborah saw his little eyes watching her, and she tried to wave, to tell him something was wrong, but she couldn't move.

At the end of the service, the congregation stood, and Deborah's mouth twisted as she fought to scream. The only sound came from Davon, who yelled, "Something's wrong with my grandma!" He sprinted from the choir platform just as Deborah fell forward onto one knee. Davon screamed, "Grandpop! Grandpop!" Pullum took one look at Deborah and yelled, "Stroke!"

The second Davon heard the word *stroke,* he grabbed Deborah's pocketbook, dug out her car keys, and ran to the car. He opened all the doors wide, laid the passenger seat back as flat as it would go, and jumped behind the wheel, feet dangling far above the pedals. Then he started the engine so Pullum could just climb in and start driving.

Soon they were speeding along the winding road from church, Deborah slipping in and out of consciousness in the passenger seat while Davon leaned over her, screaming, "Don't you go to sleep, Ma!" and slapping her hard across the face every time she closed her eyes. Pullum kept yelling for him to stop, saying, "Boy, you gonna kill your grandma!" But Davon wouldn't stop.

When they got to the fire station down the road, medics pulled Deborah from the car, gave her oxygen and injections, ran an IV to her arm, and loaded her into an ambulance. As the ambulance drove away, a fireman told Davon he was smart to smack Deborah in the car.

"Boy, you did your grandmother a favor," the fireman said. "You just saved her life."

One of the first things Deborah said when she regained consciousness was, "I have to take a test." The hospital staff thought she meant she needed a CT scan or a blood test, but she meant a test for school.

When the doctors finally let Deborah's family see her, Davon,

Pullum, and Deborah's daughter, Tonya, filed in to find Deborah sitting propped up in bed, eyes wide. Tired, but alive. Her left side was still weak, and she couldn't move her arms well, but the doctors said she was lucky and would probably recover completely.

"Praise the Lord!" Pullum yelled.

A few days later, when Deborah got out of the hospital, she left me a voice mail. It was my birthday, and we'd planned to meet in Clover that day. "Happy birthday, Boo," she said, her voice utterly calm. "I'm sorry I can't come celebrate with you down in the country, but I had me a couple strokes the other day. It was bound to happen, but praise the Lord, I'm okay. Can't talk too good out one side of my mouth, but doctor says I'm gonna be fine. You keep reportin, and don't you worry about me—I feel good. Better than since before I found out they took my mother cells. I feel so light, you know? It lifted my burden. I thank the Lord for what happened."

The doctor told Deborah a second stroke was almost always worse than the first. "Trust me," he said, "you don't want to do this again." He told her she needed to educate herself, learn the warning signs, know how to bring down her blood pressure and control her blood sugar.

"Just another reason I got to keep goin on and get to school," she told me. "I already signed up for a diabetes class and a stroke class to get more understanding about that. Maybe I can take a nutrition class to learn how to eat good, too."

The stroke seemed to ease tension in the family too: Deborah's brothers began calling every day to see how she was doing, and Zakariyya even said he wanted to visit. Deborah hoped this meant her brothers would find peace with her desire for information about their mother.

She called me laughing, saying, "Girl, I got to get my rest so we can get back on the road and do more research before the trail get cold! But from now on, I ride with you. Everything will be all right. That's what I woke up knowin. I just gotta move a little slower, pay attention to things, and not let myself get scared. Cause there's nothin to be scared about with my mother and them cells. I don't want nothin to keep me from learnin no more."

But in fact there was something that would keep Deborah from learning: she didn't have enough money. Her social security check barely covered her living expenses, let alone classes and books. She came up with several ideas for making money, including a colorful disposable baby bottle with premeasured amounts of water and formula— something a busy mom could shake with one hand while holding a baby. She drew careful diagrams and sent them off with a patent application, but she dropped the idea when she found out it would cost several thousand dollars to make the prototype.

Eventually Deborah stopped thinking about going to school herself and instead started focusing on making sure her grandchildren and grandnieces and grandnephews got educated.

"It's too late for Henrietta's children," she told me one day over the phone. "This story ain't about us anymore. It's about the new Lacks children."

Two months after Deborah's stroke, we went to Pullum's church to watch him baptize Sonny's nine-month-old baby granddaughter, JaBrea. There was hardly an empty seat when the sermon started. Pullum stood behind the pulpit wrapped in a long black robe with red crosses on its front, sweat beading his forehead. A blind piano player tapped his way to the piano and began to play as the congregation sang along: *"Stand by me, while I run this race, for I don't want to run this race in vain."*

Pullum pointed at me and grinned a mischievous grin.

"Come stand by me!" he yelled.

"Oh girl, you in trouble now," Deborah whispered, elbowing me in the ribs.

"I am *not* going up there," I whispered back. "Just pretend like we can't see him."

Pullum waved his arms over his head, then pointed at the pulpit for me to join him. Deborah and I stared at the choir behind him, our faces blank, pretending not to see. Pullum rolled his eyes, then yelled into the microphone, "We have a guest with us today! Rebecca Skloot, would you stand for us this morning?"

Deborah whispered, "Uh-oh," as the entire congregation followed his pointing finger to look at me.

I stood.

"Sister Rebecca Skloot," Pullum said, "I know this might not be the right time for you, but it's the right time for me."

"Amen," Deborah said from her seat beside me, her voice suddenly serious.

"John Hopkins took my wife's mother's body and used what they needed," he yelled into the microphone. "They sold her cells all over the world! Now I'm gonna have Sister Rebecca Skloot come and talk about what she doing with my wife and them cells."

I'd never sat in a congregation before, let alone spoken in front of one. My face flushed and my throat clenched as Deborah pushed my back to get me moving. Pullum told the congregation to give me a hand, and the room erupted in cheers. I walked to the pulpit and took the microphone from Pullum, who patted me on the back and whispered in my ear, "Just preach it in your own words." So I did. I told the story of Henrietta's cells and what they'd done for science, my voice growing louder as the congregation yelled "Amen!" and "Hallelujah!" and "Lord have mercy!"

"Most people think her name was Helen Lane," I said. "But she was Henrietta Lacks. She had five children, and one of them is sitting right over there." I pointed at Deborah. She was holding JaBrea in her lap now, grinning, tears streaming down her cheeks.

Pullum stepped forward and took the microphone, putting his arm around my shoulders and squeezing so I wouldn't walk away.

"I was very angry with Sister Rebecca when she started calling us," he said. "So was my wife. Then finally we said okay, but we told her, 'You need to talk to us like we're regular folk. You need to tell us what's goin on.' "

Then he looked at Deborah. "The world gonna know who your mother is. But you and Sonny and the rest of Henrietta's children, they probably won't see real benefits from them cells." Deborah nodded as Pullum raised his long robed arm and pointed to JaBrea, a breathtakingly beautiful baby dressed in white lace with a bow in her hair.

"This child will someday know that her great-grandmother Henrietta helped the world!" Pullum yelled. Then he pointed around the room at Davon and JaBrea's other cousins, saying, "So will that child . . . and that child . . . and that child. This is their story now. They need to take hold of it and let it teach them they can change the world too."

He raised his arms above his head and yelled hallelujah. Baby JaBrea waved her hands and let out a loud happy screech, and the congregation yelled amen.

38

The Long Road to Clover

On January 18, 2009, on a cold, sunny Sunday, I pulled off the highway onto the road into Clover. As I passed one green field after the next, I thought, *I don't remember the road into Clover being so long.* Then I realized I'd just passed the Clover post office—it was across the street from a big, empty field. *But it used to be across the street from the rest of downtown.* I didn't understand. If that was the post office, where was everything else? I kept driving for a moment, thinking, *Did they move the post office?* Then it hit me.

Clover was gone.

I jumped out of the car and ran into the field, to the spot where the old movie theater once stood—where Henrietta and Cliff once watched Buck Jones films. It was gone. So was Gregory and Martin's grocery and Abbott's clothing store. I stood with my hand over my mouth, staring in disbelief at the empty field until I realized there were splinters of brick and small white plaster tiles pressed into the dirt and grass. I knelt down and began collecting them, filling my pockets with what remained of the town of Henrietta's youth.

I have to send some of this to Deborah, I thought. *She's not going to believe Clover is gone.*

Standing on Main Street, staring at the corpse of Clover's downtown, it felt like everything related to Henrietta's history was vanishing. In 2002, just one year after Gary had wrapped his hands around Deborah's head and passed the burden of the cells on to me, he'd died suddenly at the age of fifty-two from a heart attack. He'd been walking toward Cootie's car, carrying his best suit to put in the trunk so it wouldn't get wrinkled on the way to Cootie's mother's funeral. A few months later, Deborah called to say that Cliff's brother Fred had died from throat cancer. Next it was Day, who died of a stroke, surrounded by his family. Then Cootie, who killed himself with a shotgun to the head. Each time someone died, Deborah called crying.

I thought the calls would never end.

"Death just following us and this story everywhere we go," she'd say. "But I'm hangin in there."

In the years that followed the baptism, not much changed for the Lackses. Bobbette and Lawrence went on with their lives. Lawrence didn't think about the cells much anymore, though occasionally he and Zakariyya still entertained the idea of suing Hopkins.

Sonny had a quintuple bypass in 2003, when he was fifty-six years old—the last thing he remembered before falling unconscious under the anesthesia was a doctor standing over him saying his mother's cells were one of the most important things that had ever happened to medicine. Sonny woke up more than $125,000 in debt because he didn't have health insurance to cover the surgery.

Zakariyya got kicked out of his assisted-living facility, then a Section Eight housing project, where he smashed a forty-ounce beer bottle over a woman's back and pushed her through a plate-glass window. He sometimes worked with Sonny, driving a truck.

In 2004 Deborah left her husband and moved into an assisted-living apartment of her own, which she'd longed to do for years—she was tired of fighting with Pullum, plus their row house had too many stairs. After she moved out, to cover her bills, she went to work full-time for her daughter Tonya, who'd opened an assisted-living home in her house. Each morning Deborah left the assisted-living facility where she lived, and spent the day cooking and cleaning for the five or six men living in her daughter's home. She quit after two years because her body couldn't take walking up and down stairs all day.

When Deborah officially divorced Pullum in 2006, she had to itemize her income as part of a request for the judge to waive her filing fee. She listed $732 per month from Social Security Disability and $10 per month in food stamps. Her checking account was empty.

When I went back to visit Clover and found Main Street razed, it had been a few months since Deborah and I talked. During our last call, I'd told her that the book was done, and she'd said she wanted me to come to Baltimore and read it to her, so I could talk her through the hard parts. I'd called several times since to plan the visit, but she hadn't returned my calls. I left messages, but didn't push her. *She needs some space to prepare herself,* I thought. *She'll call when she's ready.* When I got home from Clover, I called again saying, "I brought something back for you from Clover—you won't believe what's happened down there." But she didn't call back.

On May 21, 2009, after leaving many messages, I called again. Her voice-mail box was full. So I dialed Sonny's number to say something I'd said to him many times over the years: "Will you tell your sister to stop messing around and return my calls? I really need to talk to her. Our time is running out." When he answered the phone I said, "Hey Sonny, it's Rebecca," and for a moment the line went silent.

"I've been trying to find your phone number," he said, and my eyes filled with tears. I knew there was only one reason Sonny would need to call me.

Deborah had gone over to her niece's house on Mother's Day, a week and a half before my call—Sonny had made crab cakes for her, the grandchildren were there, and everyone laughed and told stories. After dinner he took Deborah back to the apartment she loved and said good night. She stayed home the next day, ate the leftover crab cakes Sonny sent home with her, and talked to Davon on the phone— he was learning to drive and wanted to come over in the morning to practice. The next morning when he called, she didn't answer. A few hours later Sonny dropped by to check on her, as he did nearly every day, and found her in her bed, arms crossed on her chest, smiling. He thought she was sleeping, so he touched her arm, saying, "Dale, time to get up." But she wasn't sleeping.

"She's in a better place now," Sonny told me. "A heart attack just after Mother's Day—she wouldn't have wanted it another way. She's suffered a lot in life, and now she's happy."

After finding Deborah in her bed, Sonny cut a lock of her hair and tucked it inside their mother's Bible with the locks of hair from Henrietta and Elsie. "She's with them now," he told me. "You know there's no place in the world she'd rather be."

Deborah was happy when she died: her grandson Little Alfred was now twelve, headed into the eighth grade, and doing well in school. Lawrence and Bobbette's granddaughter Erika had gotten into Penn State after writing an admissions essay about how her great-grandmother Henrietta's story had inspired her to study science. After transferring to the University of Maryland, she earned her bachelor's degree and entered a master's program in psychology, becoming the first of Henrietta's descendants to attend graduate school. At seventeen, Deborah's grandson Davon was about to graduate from high school. He'd promised Deborah he'd go to college and continue learning about Henrietta until he knew everything there was to know about her. "That really made me feel okay about dying whenever my time come," she'd told me.

As Sonny told me the news of Deborah's death, I sat staring at a framed picture of her that's been on my desk for nearly a decade. In it, her eyes are hard, her brow creased and angry. She's wearing a pink shirt and holding a bottle of pink Benadryl. Everything else is red: her fingernails, the welts on her face, the dirt beneath her feet.

I stared at that picture for days after her death as I listened to hours of tape of us talking, and read the notes I'd taken the last time I saw her. At one point during that visit, Deborah, Davon, and I sat side by side on her bed, our backs to the wall, legs outstretched. We'd just finished watching two of Deborah's favorite movies back-to-back: *Roots* and the animated movie *Spirit*, about a wild horse who's captured by the U.S. Army. She wanted us to watch them together so we could see the similarities between the two—Spirit fought for his freedom just as Kunta Kinte did in *Roots*, she said.

"People was always tryin to keep them down and stop them from doing what they want just like people always doin with me and the story about my mother," she said.

When the films ended, Deborah jumped out of bed and put in yet another video. She pressed PLAY and a younger version of herself appeared on the screen. It was one of nearly a dozen tapes the BBC had recorded that didn't make it into the documentary. On the screen, Deborah sat on a couch with her mother's Bible open in her lap, her hair brown instead of gray, her eyes bright, with no circles beneath them. As she talked, her hand stroked the long lock of her mother's hair.

"I often visit her hair in the Bible," Deborah said into the camera. "When I think about this hair, I'm not as lonely. I imagine, what would it be like to have a mother to go to, to laugh, cry, hug. God willing, I can be with her someday. I'm looking forward to that."

The younger Deborah said she was glad that when she died, she wouldn't have to tell her mother the story of everything that happened with the cells and the family, because Henrietta already knew. "She's been watching us and seeing all that's going on down here," Deborah said. "She's waiting patiently for us. There won't be any words, just a lot of hugging and crying. I really believe she's up in heaven, and she's doin okay, because she did enough suffering for everyone down here. On the other side, they say there's no pain or suffering. . . . I want to be there with my mother."

Sitting between me and Davon on the bed, Deborah nodded at her younger self on the screen and said, "Heaven looks just like Clover,

Virginia. My mother and I always loved it down there more than anywhere else in the world."

She stroked Davon's hair. "I don't know how I'm going to go," she said. "I just hope it's nice and calm. But I tell you one thing, I don't want to be immortal if it mean living forever, cause then everybody else just die and get old in front of you while you stay the same, and that's just sad." Then she smiled. "But maybe I'll come back as some HeLa cells like my mother, that way we can do good together out there in the world." She paused and nodded again. "I think I'd like that."

Where They Are Now

Alfred Carter Jr., Deborah's son, is in prison, serving a thirty-year sentence for robbery with a dangerous and deadly weapon, and first-degree assault with a handgun. While incarcerated, he went through drug and alcohol rehabilitation, got his GED, and taught GED classes to other inmates for twenty-five dollars a month. In 2006 he wrote to the judge who sentenced him, saying he wanted to pay back the money he stole and needed to know who to send it to.

Dr. Sir Lord Keenan Kester Cofield's whereabouts are unknown. Most recently, he served several years in prison for trying to buy jewelry at Macy's with a stolen check, and filed several lawsuits while incarcerated. In 2008, after being released from prison, Cofield filed a seventy-five-page lawsuit—his last to date—that a judge called "incomprehensible." He sued 226 parties for more than $10 billion, and argued that past decisions in all his cases should be reversed in his favor, and that anyone who'd printed his name without permission should be included in his suit, because he'd copyrighted his name. I was never able to get in touch with him to interview him for this book.

Cliff Garret, Henrietta's cousin, lived in his farmhouse in Clover until 2009, when his failing health required him to move in with his son in Richmond, Virginia, where he currently lives.

HeLa is still one of the most commonly used cell lines in laboratories around the world. When this book went to press in 2009, more than 60,000 scientific articles had been published about research done on HeLa, and that number was increasing steadily at a rate of more than 300 papers each month. HeLa cells are still contaminating other cultures and causing an estimated several million dollars in damage each year.

Howard Jones, Henrietta's doctor, is an emeritus professor at Johns Hopkins and Eastern Virginia Medical School. He founded the Jones Institute for Reproductive Medicine in Norfolk, Virginia, with his late wife, Georgeanna. They were pioneers in the field of infertility treatments, and were responsible for the first test-tube baby born in the United States. When this book went to press, he was ninety-nine years old.

Mary Kubicek is retired and living in Maryland.

Zakariyya, Sonny, and **Lawrence Lacks** were deeply affected by Deborah's death. Lawrence charged more than $6,000 to his credit cards to cover the cost of her burial, and when this book went to press, Sonny was saving money to buy her a tombstone. Zakariyya stopped drinking and began studying the lives of yogis and others who'd achieved inner peace. He started spending more time with his family, including his many nieces and nephews, who hug and kiss him on a regular basis. He smiles often. Sonny has sworn to uphold Deborah's desire to gain recognition for their mother. Today, when the Lacks brothers talk about Henrietta, they focus on the importance of her contribution to science. They no longer talk about suing Johns Hopkins, though Lawrence and Zakariyya still believe they're owed a share of the profits from HeLa cells.

Christoph Lengauer is Global Head of Oncology Drug Discovery at sanofi-aventis, one of the biggest pharmaceutical companies in the world. Many of the scientists working for him use HeLa cells on a routine basis. He lives in Paris, France.

Davon Meade and **(Little) Alfred Jr.**, Deborah's grandsons, live in Baltimore, as do twenty-two of Henrietta's other descendants, including her grandchildren, great-grandchildren, and great-great-grandchildren. Two others live in California.

John Moore appealed to the U.S. Supreme Court, which refused to hear his case. He died in 2001.

Roland Pattillo is a professor at Morehouse School of Medicine, where he continues to hold his HeLa conference in Henrietta's honor each year. Pattillo and his wife, Pat, bought a marker for Henrietta's grave.

James Pullum, Deborah's ex-husband, is still preaching in Baltimore.

Courtney Speed still runs her grocery store, where she continues to teach local children to do math, and hopes to open a Henrietta Lacks Museum.

About the
Henrietta Lacks
Foundation

Before *The Immortal Life of Henrietta Lacks* was published, Rebecca Skloot established the Henrietta Lacks Foundation. Some of the proceeds of this book are being donated by the author to the Foundation. Visit HenriettaLacksFoundation.org for more information.

Afterword

When I tell people the story of Henrietta Lacks and her cells, their first question is usually *Wasn't it illegal for doctors to take Henrietta's cells without her knowledge? Don't doctors have to tell you when they use your cells in research?* The answer is no—not in 1951, and not in 2009, when this book went to press.

Today most Americans have their tissue on file somewhere. When you go to the doctor for a routine blood test or to have a mole removed, when you have an appendectomy, tonsillectomy, or any other kind of *ectomy*, the stuff you leave behind doesn't always get thrown out. Doctors, hospitals, and laboratories keep it. Often indefinitely.

In 1999 the RAND Corporation published a report (the first and, so far, last of its kind) with a "conservative estimate" that more than 307 million tissue samples from more than 178 million people were stored in the United States alone. This number, the report said, was increasing by more than 20 million samples each year. The samples come from routine medical procedures, tests, operations, clinical trials, and research donations. They sit in lab freezers, on shelves, or in industrial vats of liquid nitrogen. They're stored at military facilities, the FBI, and the National Institutes of Health. They're in biotech company labs and most hospitals. Biobanks store appendixes, ovaries, skin, sphincters, testicles, fat, even foreskins from most circumcisions. They also house blood samples taken from most infants born in the United States since the late sixties, when states started mandating the screening of all newborns for genetic diseases.

And the scale of tissue research is only getting bigger. "It used to be, some researcher in Florida had sixty samples in his freezer, then another guy in Utah had some in his," says Kathy Hudson, a molecular biologist who founded the Genetics and Public Policy Center at Johns Hopkins University and is now chief of staff at NIH. "Now we're talking about a massive, massive scale." In 2009 the NIH invested $13.5 million to develop a bank for the samples taken from newborns nationwide. A few years ago the National Cancer Institute started gathering what it expects will be millions of tissue samples for mapping cancer genes; the Genographic Project began doing the same to map human migration patterns, as did the NIH to track disease genes. And for several years the public has been sending samples by the millions to personalized DNA testing companies like 23andMe, which only provide customers with their personal medical or genealogical information if they first sign a form granting permission for their samples to be stored for future research.

Scientists use these samples to develop everything from flu vaccines to penis-enlargement products. They put cells in culture dishes and expose them to radiation, drugs, cosmetics, viruses, household chemicals, and biological weapons, and then study their responses. Without those tissues, we would have no tests for diseases like hepatitis and HIV; no vaccines for rabies, smallpox, measles; none of the promising new drugs for leukemia, breast cancer, colon cancer. And developers of the products that rely on human biological materials would be out billions of dollars.

How you should feel about all this isn't obvious. It's not as if scientists are stealing your arm or some vital organ. They're using tissue scraps you parted with voluntarily. Still, that often involves someone taking part of you. And people often have a strong sense of ownership when it comes to their bodies. Even tiny scraps of them. Especially when they hear that someone else might be making money off those scraps, or using them to uncover potentially damaging information about their genes and medical histories. But a *feeling* of ownership doesn't hold up in court. And at this point no case law has fully clarified whether you own or have the right to control your tissues. When they're part of your body, they're clearly yours. Once they're excised, your rights get murky.

Kathy Hudson, who has conducted focus groups about the public's feelings on the tissue issue, says she believes that tissue rights have the potential to become a bona fide movement.

"I could see people starting to say, 'No, you can't take my tissues,'" she told me. "All I can say is, we better deal with the problems now instead of waiting until that happens."

There are, essentially, two issues to deal with: consent and money. For most people, knowing if and how their tissues are being used in research is a far bigger issue than profiting from them. Yet when this book went to press, storing blood and tissues for research did not legally require informed consent, because the law governing such things doesn't generally apply to tissue research.

The Federal Policy for the Protection of Human Subjects, also known as the Common Rule, requires informed consent for all human-subject research. But in practice, most tissue research isn't covered because: (1) it's not federally funded, or (2) the researcher never learns the identity of the "donors" or has firsthand contact with them, in which case it's not considered research on humans. So in the end, the Common Rule doesn't actually govern most tissue research.

Today, if doctors want to gather tissues from patients strictly for research purposes—as in Henrietta's case—they are required to get informed consent. But storing tissues from diagnostic procedures like, say, mole biopsies, and using them in future research doesn't require such consent. Most institutions still choose to get permission, but there's no uniformity in the way that's done. A few hand out enough information to fill a small book, explaining exactly what will be done with all patient tissues. But most just include a short line in an admission form saying that any tissues removed may be used for education or research.

According to Judith Greenberg, director of the Division of Genetics and Developmental Biology at the National Institute of General Medical Science, the NIH now has "very stringent guidelines" requiring consent for any tissues collected for their banks. "It's very important for donors to understand what the consequences of tissue research might be," she says. But their guidelines apply only to NIH research, and they're not legally binding.

Supporters of the status quo argue that passing new, tissue-specific legislation is unnecessary, and that the current oversight practices are enough. They point to Institutional Review Boards; the many professional guidelines, like the American Medical Association's Code of Ethics (which requires doctors to inform patients if their tissue samples will be used in research or lead to profits); and several post Nuremberg codes, including the Declaration of Helsinki and the Belmont Report, all of which list consent as a requirement. But guidelines and ethical codes aren't laws, and many tissue-rights supporters say internal review doesn't work.

Beyond simply knowing their tissues are being used in research, some tissue-rights activists believe donors should have the right to say, for example, that they don't want their tissues used for research on nuclear weapons, abortion, racial differences, intelligence, or anything else that might run contrary to their beliefs. They also believe it's important for donors to be able to control who has access to their tissues, because they worry that information gathered from tissue samples might be used against them.

In 2005, members of the Native American Havasupai Tribe sued

Arizona State University after scientists took tissue samples the tribe donated for diabetes research and used them without consent to study schizophrenia and inbreeding. Their case is still pending. In 2006, some seven hundred new mothers found out that doctors had taken their placentas without consent to test for abnormalities that might help the hospital defend itself against future lawsuits over birth defects. And in a handful of cases, genetic tests performed on people without their consent have been used to deny workers' compensation or health insurance claims (something now protected against by the Genetic Information Nondiscrimination Act of 2008).

Because of cases like these, a growing number of activists—ethicists, lawyers, doctors, and patients—are arguing cases and pushing for new regulations that would grant people the right to control their tissues. And a growing number of tissue "donors" are suing for control of their samples and the DNA inside them. In 2005, six thousand patients demanded that Washington University remove their tissue samples from its prostate-cancer bank. The university refused, and the samples were tied up in litigation for years. So far, two courts have ruled against the patients, relying on the same logic used in the Moore case (that giving patients those rights would inhibit research, etc). In 2008 the patients appealed to the Supreme Court, which refused to hear their case. When this book went to press, they were contemplating class action. Most recently, in July 2009, parents in Minnesota and Texas sued to stop the nationwide practice of storing and conducting research—without consent—on fetal blood samples, many of which can be traced back to the infants they came from. They argue that research on those samples is an invasion of their children's privacy.

Because of the Health Insurance Portability and Accountability Act (HIPAA) of 1996, there is now clear federal law in place to prevent the kind of privacy violation that happened to the Lacks family when doctors at Hopkins released Henrietta's name and her medical records. Since tissues connected to their donors' names are subject to strict regulation under the Common Rule, samples are no longer named using donor initials as Henrietta's cells were; today they're usually identified by code numbers. But, as Judith Greenberg of the NIH says, "It's never possible to one-hundred-percent guarantee anonymity, because in theory we can now sequence genes and find out who anyone is from their cells. So the consent process has to be more about laying out the risks of tissue research so people can decide whether they want to participate."

Ellen Wright Clayton, a physician and lawyer who is director of the Center for Biomedical Ethics and Society at Vanderbilt University, says there needs to be a "very public conversation" about all of this. She says, "If someone presented a bill in Congress that said, 'As of today, when you go to the doctor for health care, your medical records and tissue samples can be used for research and nobody has to ask you'—if the issue were stated that bluntly so people could really understand what's happening and say they're okay with it, that would make me more comfortable with what we're currently doing. Because what's happening now is not what people think is going on."

Lori Andrews, director of the Institute for Science, Law and Technology at the Illinois Institute of Technology, wants something more drastic: she has called for people to get policymakers' attention by becoming "conscientious objectors in the DNA draft" and refusing to give tissue samples.

David Korn, vice provost for research at Harvard University, argues that giving patients control over their tissues is shortsighted. "Sure," he says, "consent feels nice. Letting people decide what's going to happen with their tissue seems like the right thing to do. But consent diminishes the value of tissue." To illustrate this, Korn points to the Spanish flu pandemic. In the 1990s, scientists used stored tissue samples from a soldier who died in 1918 to re-create the virus's genome and study why it was so deadly, with hopes of uncovering information about the current avian flu. In 1918, asking that soldier's permission to take tissues for this kind of future research would have been impossible, Korn says. "It was an inconceivable question—no one even knew what DNA was!"

For Korn, the consent issue is overshadowed by a public responsibility to science: "I think people are morally obligated to allow their bits and pieces to be used to advance knowledge to help others. Since everybody benefits, everybody can accept the small risks of having their tissue scraps used in research." The only exception he would make is for people whose religious beliefs prohibit tissue donation. "If somebody says being buried without all their pieces will condemn them to wandering forever because they can't get salvation, that's legitimate, and people should respect it," Korn says. But he acknowledges that people can't raise those objections if they don't understand their tissues are being used in the first place.

"Science is not the highest value in society," Andrews says, pointing instead to things like autonomy and personal freedom. "Think about it," she says. "I decide who gets my money after I die. It wouldn't harm *me* if I died and you gave all my money to someone else. But there's something psychologically beneficial to me as a living person to know I can give my money to whoever I want. No one can say, 'She shouldn't be allowed to do *that* with her money because *that* might not be most beneficial to society.' But replace the word *money* in that sentence with *tissue,* and you've got precisely the logic many people use to argue against giving donors any control over their tissues."

Wayne Grody, director of the Diagnostic Molecular Pathology Laboratory at the University of California, Los Angeles, was once a fierce opponent of consent for tissue research. But after years of debating people like Andrews and Clayton, he's become more moderate. "I'm pretty convinced that we should go the extra mile to have a good and complex consent process," he told me. Still, he can't imagine how it would work. "These tissues enter a pipeline of millions of other samples," he said. "How are you going to distinguish, well, this patient said we can study colon cancer; the next one said we can do anything we want, but we can't commercialize it. I mean, do they all have to be color-coded?" Regardless, Grody stresses that questions of consent should apply only to the collection of future samples, not the millions already stored, including HeLa. "What are we going to do," he says, "throw them all out?"

If the issue of consent isn't addressed, Robert Weir, founder of the biomedical ethics center at the University of Iowa, sees only one outcome: "Patients turn to law as a last resort when they don't see their participation being acknowledged." Weir favors fewer lawsuits and more disclosure. "Let's get these things on the table and come up with legal guidelines we can all live with," he says. "Because going to court is the only other option." And court is where these cases often end up, particularly when they involve money.

When it comes to money, the question isn't *whether* human tissues and tissue research will be commercialized. They are and will continue to be; without commercialization, companies wouldn't make the drugs and diagnostic tests so many of us depend on. The question is how to deal with this commercialization—whether scientists should be required to tell people their tissues may be used for profit, and where the people who donate those raw materials fit into that marketplace.

It's illegal to sell human organs and tissues for transplants or medical treatments, but it's perfectly legal to give them away while charging fees for collecting and processing them. Industry-specific figures don't exist, but estimates say one human body can bring in anywhere from $10,000 to nearly $150,000. But it's extremely rare for individual cells from one person to be worth millions like John Moore's. In fact, just as one mouse or one fruit fly isn't terribly useful for research, most individual cell lines and tissue samples aren't worth anything on their own. Their value for science comes from being part of a larger collection.

Today, tissue-supply companies range from small private businesses to huge corporations, like Ardais, which pays the Beth Israel Deaconess Medical Center, Duke University Medical Center, and many others an undisclosed amount of money for exclusive access to tissues collected from their patients.

"You can't ignore this issue of who gets the money and what the money is used for," says Clayton. "I'm not sure what to do about it, but I'm pretty sure it's weird to say everybody gets money except the people providing the raw material."

Various policy analysts, scientists, philosophers, and ethicists have suggested ways to compensate tissue donors: creating a Social Security–like system in which each donation entitles a person to increasing levels of compensation; giving donors tax write-offs; developing a royalty system like the one used for compensating musicians when their songs are played on the radio; requiring that a percentage of profits from tissue research go to scientific or medical charities, or that all of it be funneled back into research.

Experts on both sides of the debate worry that compensating patients would lead to profit-seekers inhibiting science by insisting on unrealistic financial agreements or demanding money for tissues used in noncommercial or nonprofit research. But in the majority of cases, tissue donors haven't gone after profits at all. They, like most tissue-rights activists, are less concerned about personal profits than about making sure the knowledge scientists gain by studying tissues is available to the public, and to other researchers. In fact, several patient groups have created their own tissue banks so they can control the use of their tissues and the patenting of discoveries related to them, and one woman became a patent holder on the disease gene discovered in her children's tissues, which lets her determine what research is done on it and how it's licensed.

Gene patents are the point of greatest concern in the debate over ownership of human biological materials, and how that ownership might interfere with science. As of 2005 — the most recent year figures were available — the U.S. government had issued patents relating to the use of about 20 percent of known human genes, including genes for Alzheimer's, asthma, colon cancer, and, most famously, breast cancer. This means pharmaceutical companies, scientists, and universities control what research can be done on those genes, and how much resulting therapies and diagnostic tests will cost. And some enforce their patents aggressively: Myriad Genetics, which holds the patents on the

BRCA1 and BRCA2 genes responsible for most cases of hereditary breast and ovarian cancer, charges $3,000 to test for the genes. Myriad has been accused of creating a monopoly, since no one else can offer the test, and researchers can't develop cheaper tests or new therapies without getting permission from Myriad and paying steep licensing fees. Scientists who've gone ahead with research involving the breast-cancer genes without Myriad's permission have found themselves on the receiving end of cease-and-desist letters and threats of litigation.

In May 2009 the American Civil Liberties Union, several breast-cancer survivors, and professional groups representing more than 150,000 scientists sued Myriad Genetics over its breast-cancer gene patents. Among other things, scientists involved in the case claim that the practice of gene patenting has inhibited their research, and they aim to stop it. The presence of so many scientists in the suit, many of them from top institutions, challenges the standard argument that ruling against biological patents would interfere with scientific progress.

Lori Andrews, who has worked pro bono on all of the most important biological ownership cases to date, including the current breast cancer gene suit, says that many scientists have interfered with science in precisely the way courts always worried tissue donors might do. "It's ironic," she told me. "The Moore court's concern was, if you give a person property rights in their tissues, it would slow down research because people might withhold access for money. But the Moore decision backfired—it just handed that commercial value to researchers." According to Andrews and a dissenting California Supreme Court judge, the ruling didn't prevent commercialization; it just took patients out of the equation and emboldened scientists to commodify tissues in increasing numbers. Andrews and many others have argued that this makes scientists less likely to share samples and results, which slows research; they also worry that it interferes with health-care delivery.

There is some evidence to support their claim. One survey found that 53 percent of laboratories had stopped offering or developing at least one genetic test because of patent enforcement, and 67 percent felt patents interfered with medical research. Because of patent licensing fees, it costs $25,000 for an academic institution to license the gene for researching a common blood disorder, hereditary haemochromatosis, and up to $250,000 to license the same gene for commercial testing. At that rate, it would cost anywhere from $46.4 million (for academic institutions) to $464 million (for commercial labs) to test one person for all known genetic diseases.

The debate over the commercialization of human biological materials always comes back to one fundamental point: like it or not, we live in a market-driven society, and science is part of that market. Baruch Blumberg, the Nobel Prize–winning researcher who used Ted Slavin's antibodies for hepatitis B research, told me, "Whether you think the commercialization of medical research is good or bad depends on how into capitalism you are." On the whole, Blumberg said, commercialization is good; how else would we get the drugs and diagnostic tests we need? Still, he sees a downside. "I think it's fair to say it's interfered with science," he said. "It's changed the spirit." Now there are patents and proprietary information where there once was free information flow. "Researchers have become entrepreneurs. That's boomed our economy and created incentives to do research. But it's also brought problems, like secrecy and arguments over who owns what."

Slavin and Blumberg never used consent forms or ownership-transfer agreements; Slavin just held up his arm and gave samples. "We lived in a different ethical and commercial age," Blumberg said. He imagines patients might be less likely to donate now: "They probably want to maximize their commercial possibilities just like everyone else."

All the important science Blumberg has done over the years depended on free and unlimited access to tissues. But Blumberg says he doesn't think keeping patients in the dark is the way to get that access: "For somebody like Ted who really needed that money to survive, it would have been wrong to say scientists could commercialize those antibodies, but he couldn't. You know, if someone was going to make money off his antibodies, why shouldn't he have a say in that?"

Many scientists I've talked to about this issue agree. "This is a capitalist society," says Wayne Grody. "People like Ted Slavin took advantage of that. You know, the way I see it is, if you think of doing that on the front end, more power to you."

The thing is, people can't "think of doing that on the front end" unless they know their tissues might be valuable to researchers in the first place. The difference between Ted Slavin, John Moore, and Henrietta Lacks was that someone told Slavin his tissues were special and that scientists would want to use them in research, so he was able to control his tissues by establishing his terms *before* anything left his body. In other words, he was informed, and he gave consent. In the end, the question is how much science should be obligated (ethically and legally) to put people in the position to do the same as Slavin. Which brings us back to the complicated issue of consent.

Just as there is no law requiring informed consent for storing tissues for research, there is no clear requirement for telling donors when their tissues might result in profits. In 2006 an NIH researcher gave thousands of tissue samples to the pharmaceutical company Pfizer in exchange for about half a million dollars. He was charged with violating a federal conflict of interest law, not because he failed to disclose his financial interest or the value of those tissues to the donors, but because federal researchers aren't allowed to take money from pharmaceutical companies. His case resulted in a congressional investigation and later a hearing; the possible interests of the patients, and their lack of knowledge of the value of their samples, wasn't mentioned at any point in the process.

Though the judge in the John Moore case said patients must be told if their tissues have commercial potential, there was no law enacted to enforce that ruling, so it remains only case law. Today the decision to disclose this information is up to the institution, and many choose not to tell patients. Some consent forms don't mention money at all; others come right out and say, "We may give or sell the specimen and certain medical information about you." Others simply say, "You will receive no reimbursement for donating tissue." Still others embrace confusion: "Your sample will be owned by [the university]. . . . It is unknown whether you will be able to gain (participate in) any financial compensation (payment) from any benefits gained from this research."

Tissue-rights activists argue that it's essential to disclose any potential financial gain that might come from people's tissues. "This isn't about trying to get patients a cut of the financial action," says Lori Andrews. "It's about allowing people to express their desires." Clayton agrees, but says, "The fundamental problem here isn't the money; it's the notion that the people these tissues come from don't matter."

After the Moore case, Congress held hearings and commissioned reports that uncovered the millions of dollars being made from human tissue research, and it formed a special committee to assess the situation and recommend how to proceed. Its findings: the use of human cells and tissues in biotechnology holds "great promise" for improving human health, but raises extensive ethical and legal questions that "have not been answered" and to which "no single body of law, policy or ethics applies." This, they said, must be clarified.

In 1999, President Clinton's National Bioethics Advisory Commission (NBAC) issued a report saying that federal oversight of tissue research was "inadequate" and "ambiguous." It recommended specific changes that would ensure patients' rights to control how their tissues were used. It skirted the issue of who should profit from the human body, saying simply that the issue "raises a number of concerns," and should be investigated further. But little happened.

Years later, I asked Wayne Grody, who was in the thick of the debate in the nineties, why the congressional recommendations and NBAC report seemed to have vanished.

"It's weird, but I have no idea," he said. "If you can figure that out, I'd like to know. We all just wanted to forget about it, like if we ignored it, maybe it would just go away." But it didn't. And given the steady flow of court cases related to tissues, the issue isn't going away anytime soon.

Despite all the other cases and the press they've received, the

Lacks family has never actually tried to sue anyone over the HeLa cells. Several lawyers and ethicists have suggested to me that since there is no way to anonymize HeLa cells at this point, research on them should be covered by the Common Rule. And since some of the DNA present in Henrietta's cells is also present in her children, it's possible to argue that by doing research on HeLa, scientists are also doing research on the Lacks children. Since the Common Rule says that research subjects must be allowed to withdraw from research at any time, these experts have told me that, in theory, the Lacks family might be able to withdraw HeLa cells from all research worldwide. And in fact, there are precedents for such a case, including one in which a woman successfully had her father's DNA removed from a database in Iceland. Every researcher I've mentioned that idea to shudders at the thought of it. Vincent Racaniello, a professor of microbiology and immunology at Columbia University, who once calculated that he's grown about 800 billion HeLa cells for his own research, says that restricting HeLa cell use would be disastrous. "The impact that would have on science is inconceivable," he said.

As for the Lackses, they have few legal options. They couldn't sue over the cells being taken in the first place for several reasons, including the fact that the statute of limitations passed decades ago. They could attempt to stop HeLa research through a lawsuit, arguing that it's impossible to anonymize Henrietta's cells, which contain their DNA. But many legal experts I've talked with doubt such a case would succeed. Regardless, the Lackses aren't interested in stopping all HeLa research. "I don't want to cause problems for science," Sonny told me as this book went to press. "Dale wouldn't want that. And besides, I'm proud of my mother and what she done for science. I just hope Hopkins and some of the other folks who benefited off her cells will do something to honor her and make right with the family."

Cast of Characters

IMMEDIATE LACKS FAMILY

David "Day" Lacks—Henrietta's husband and cousin

David Jr. "Sonny" Lacks—Henrietta and Day's third child

Deborah "Dale" Lacks—Henrietta and Day's fourth child

Elsie Lacks (born Lucille Elsie Pleasant)—Henrietta's second born and eldest daughter. She was institutionalized due to epilepsy and died at age 15.

Eliza Lacks Pleasant—Henrietta's mother. She died when Henrietta was four.

Gladys Lacks—Henrietta's sister, who disapproved of Henrietta's marriage to Day

Johnny Pleasant—Henrietta's father. He left his ten children when their mother died.

Lawrence Lacks—Henrietta and Day's firstborn child

Loretta Pleasant—Henrietta's birth name

Tommy Lacks—Henrietta and Day's grandfather who raised both of them

Zakariyya Bari Abdul Rahman (born Joe Lacks)—Henrietta and Day's fifth child. Henrietta was diagnosed with cervical cancer shortly after his birth.

EXTENDED LACKS FAMILY

Albert Lacks—Henrietta's white great-grandfather. He had five children by a former slave named Maria and left part of the Lacks plantation to them. This section became known as "Lacks Town."

Alfred "Cheetah" Carter—Deborah's first husband. The marriage was abusive and ended in divorce.

Alfred Jr.—Deborah and Cheetah's firstborn child and Little Alfred's father

Bobette Lacks—Lawrence's wife. She helped raise Lawrence's siblings after Henrietta's death, and advocated for them when she discovered they were being abused.

Cliff Garret—Henrietta's cousin. As children, they worked the tobacco fields together.

"Crazy Joe" Grinnan—Henrietta's cousin who competed unsuccessfully with Day for her affection

Davon Meade—Deborah's grandson who often lived with and took care of her

Ethel—Galen's wife, an abusive caregiver to Henrietta's three youngest children

Fred Garret—Henrietta's cousin who convinced Day and Henrietta to move to Turner Station

Galen—Henrietta's cousin. He and his wife, Ethel, moved in with Day after Henrietta's death to help take care of the children. He ended up abusing Deborah.

Gary Lacks—Gladys's son and Deborah's cousin. A lay preacher, he performed a faith healing on Deborah.

LaTonya—Deborah and Cheetah's second child; Davon's mother

"Little Alfred"—Deborah's grandson

Margaret Sturdivant—Henrietta's cousin and confidante. Henrietta went to her house after radiation treatments at Johns Hopkins.

Reverend James Pullum—Deborah's second ex-husband, a former steel-mill worker who became a preacher

Sadie Sturdivant—Margaret's sister, Henrietta's cousin and confidante, she supported Henrietta during her illness. She and Henrietta sometimes sneaked out to go dancing.

MEMBERS OF THE MEDICAL AND SCIENTIFIC COMMUNITY

Alexis Carrel—French surgeon and Nobel Prize recipient who claimed to have cultured "immortal" chicken-heart cells

Chester Southam—cancer researcher who conducted unethical experiments to see whether or not HeLa could "infect" people with cancer

Christoph Lengauer—cancer researcher at Johns Hopkins who helped develop FISH, a technique used to detect and identify DNA sequences, and who reached out to members of the Lacks family

Emanuel Mandel—director of medicine at the Jewish Chronic Disease Hospital (JCDH) who partnered with Southam in unethical experiments

Dr. George Gey—head of tissue-culture research at Johns Hopkins. He developed the techniques used to grow HeLa cells from Henrietta's cancer tissue in his lab.

Howard Jones—Henrietta's gynecologist at Johns Hopkins

Leonard Hayflick—Microbiologist who proved that normal cells die when they've doubled about fifty times. This is known as the Hayflick limit.

Margaret Gey—George Gey's wife and research assistant. She was trained as a surgical nurse.

Mary Kubicek—George Gey's lab assistant who cultured HeLa cells for the first time

Richard Wesley TeLinde—one of the top cervical cancer experts in the country at the time of Henrietta's diagnosis. His research involved taking tissue samples from Henrietta and other cervical cancer patients at Johns Hopkins.

Roland Pattillo—professor of gynecology at Morehouse School of Medicine who was one of George Gey's only African-American students. He organizes a yearly HeLa conference at Morehouse in Henrietta's honor.

Stanley Gartler—the geneticist who dropped the "HeLa bomb" when he proposed that many of the most commonly used cell cultures had been contaminated by HeLa

Susan Hsu—the postdoctoral student in Victor McKusick's lab who was assigned to make contact with the Lackses and request samples from them for genetic testing without informed consent

Victor McKusick—geneticist at Johns Hopkins who conducted research on samples taken from Henrietta's children without informed consent to learn more about HeLa cells

Walter Nelson-Rees—the geneticist who tracked and published the names of cell lines contaminated with HeLa without first warning the researchers he exposed. He became known as a vigilante.

Journalists and Others

Courtney "Mama" Speed—resident of Turner Station and owner of Speed's Grocery. She organized an effort to build a Henrietta Lacks museum.

John Moore—cancer patient who unsuccessfully sued his doctor and the regents of the University of California over the use of his cells to create the Mo cell line

Michael Gold—author of *A Conspiracy of Cells.* He published details from Henrietta's medical records and autopsy report without permission from the Lacks family.

Michael Rogers—*Rolling Stone* reporter who wrote an article about the Lacks family in 1976. He was the first journalist to contact the Lackses.

Sir Lord Keenan Kester Cofield—attempted to sue Johns Hopkins and the Lacks family

Ted Slavin—a hemophiliac whose doctor told him his cells were valuable. Slavin founded Essential Biologicals, a company that sold his cells, and later cells from other people so individuals could profit from their own biological materials.

1889 Johns Hopkins Hospital is founded.

1912 Alexis Carrel claims to have successfully grown "immortal" chicken-heart cells.

1920 Henrietta Lacks is born in Roanoke, Virginia.

1947 The Nuremberg Code, a set of ethical standards for human experimentation, is produced as the result of a trial against several Nazi doctors who conducted experiments on prisoners during WWII.

1951 George Gey successfully cultures the first immortal human cell line using cells from Henrietta's cervix. It is given the name HeLa after the first two initials of Henrietta's first and last names.

1951 Henrietta Lacks dies of an unusually aggressive strain of cervical cancer.

1952 HeLa cells become the first living cells shipped via postal mail.

1952 The Tuskegee Institute opens the first "HeLa factory," supplying cells to laboratories and researchers and operating as a nonprofit. Within a few years, a company named Microbiological Associates would begin selling HeLa for profit.

1952 Scientists use HeLa cells to help develop the polio vaccine.

1953 HeLa cells become the first cells ever cloned.

1954 The pseudonym "Helen Lane" first appears in print as the source of HeLa cells.

1954 Chester Southam begins to conduct experiments without patient consent to see whether or not injections of HeLa cells could cause cancer.

1957 The term "informed consent" first appears in court documents.

1965 HeLa cells are fused with mouse cells, creating the first animal-human hybrid cells.

1965 The Board of Regents of the University of the State of New York finds Southam and a colleague guilty of unprofessional conduct and calls for stricter guidelines regarding human research subjects and informed consent.

1966 To ensure adherence to the new guidelines for research involving human subjects, the National Institutes of Health begins requiring the approval of Institutional Review Boards for any research they fund.

1966 Stanley Gartler drops the "HeLa Bomb" and proposes that HeLa cells have contaminated numerous cell lines.

1970 George Gey dies of pancreatic cancer.

1971 In a tribute to Gey, Henrietta Lacks is correctly identified for the first time in print as the source of HeLa.

1973 The Lacks family learns for the first time that Henrietta's cells are still alive.

1973 Researchers from Johns Hopkins take samples from Henrietta's children to further HeLa research, without informed consent.

1974 The Federal Policy for the Protection of Human Subjects (the Common Rule) requires informed consent for all human-subject research.

1975 Michael Rogers publishes an article about HeLa and the Lacks family in *Rolling Stone*. The Lacks family learns for the first time that Henrietta's cells have been commercialized.

1984 John Moore unsuccessfully sues his doctor and the Board of Regents of the University of California for property rights over his tissues. Moore appeals the decision.

1985 Portions of Henrietta's medical records are published without her family's knowledge or consent.

1988 The California Court of Appeals rules in John Moore's favor, saying that patients must have the power to control what becomes of their own tissues. Moore's doctor and the University of California appeal.

1991 The Supreme Court of California rules against John Moore, saying that once tissues are removed from the body, with or without consent, a person no longer owns those tissues.

1996 The Health Insurance Portability and Accountability Act makes it illegal for healthcare providers or health insurers to make personal medical information public.

1999 The RAND Corporation publishes a report with a "conservative estimate" that more than 307 million tissue samples from more than 178 million people are stored in the United States alone. The majority of the samples were taken without consent.

2005 Members of the Native American Havasupai tribe sue Arizona State University after scientists take tissue samples the tribe donated for diabetes research and use them without consent to study schizophrenia and inbreeding.

2005 Six thousand patients join a lawsuit against Washington University, demanding that the university remove their tissue samples from its prostate-cancer bank. Two courts later rule against the patients.

2005 By this date, the U.S. government has issued patents relating to the use of about 20 percent of known human genes, including genes for Alzheimer's, asthma, colon cancer, and, most famously, breast cancer.

2006 An NIH researcher is charged with violating a federal conflict-of-interest law for providing thousands of tissue samples to the pharmaceutical company Pfizer in exchange for about half a million dollars.

2009   The National Institutes of Health invests $13.5 million to develop a bank for fetal blood samples.

2009   Parents in Minnesota and Texas sue to stop the nationwide practice of storing and conducting research—without consent—on fetal blood samples, many of which can be traced back to the infants they came from.

2009   More than 150,000 scientists join the American Civil Liberties Union and breast-cancer patients in suing Myriad Genetics over its breast-cancer gene patents. The suit claims that the practice of gene patenting violates patent law and has inhibited scientific research.

Acknowledgments

Time and time again, I saw people energized by the story of Henrietta and her cells—energized, and filled with the desire to do something to show their thanks for her contribution to science, and make amends to her family. Many of those people put that energy into helping me with this book. My gratitude goes out to everyone who devoted time, knowledge, money, and heart to this project. I do not have room to name all of you here, but I could not have written this book without you.

First and foremost, I owe endless thanks to Henrietta Lacks's family.

Deborah was the soul of this book—her spirit, her laughter, her pain, her determination, and her unbelievable strength were an inspiration that helped keep me working all these years. I feel deeply honored to have been part of her life.

I thank Lawrence and Zakariyya for their trust and their stories, and Sonny, for seeing the value of this project and being its backbone within the family. I thank him for his honesty, his never-ending optimism, and for believing I could and would write this book.

Deborah's grandsons, Davon and Alfred, were incredibly supportive of Deborah's quest to learn about her mother and her sister. I thank them for keeping us laughing and for answering my many questions. Bobbette Lacks, a strong woman who has helped hold the Lacks family together for decades, put up with hours of interviews and many requests for documents, and she never held back when it came to sharing her stories. I'm grateful to Sonny's ever-reliable daughter, Jeri Lacks-Whye, who tracked down facts and photos, and often wrangled her big extended family on my behalf. I thank her and her mother, Shirley Lacks, as well as Lawrence's granddaughters Erika Johnson and Courtnee Simone Lacks, and Deborah's son, Alfred Carter Jr., for their openness and enthusiasm. James Pullum's support was unwavering; I thank him for his stories, his laughter, and his prayers. The same is true for

Gary Lacks, who sang beautiful hymns into my telephone voice mail, and never failed to serenade me on my birthday.

Re-creating the life of Henrietta Lacks wouldn't have been possible without the generous help of her family, friends, and neighbors, particularly Fred Garret, Howard Grinnan, Hector "Cootie" Henry, Ben Lacks, Carlton Lacks, David "Day" Lacks Sr., Emmett Lacks, Georgia Lacks, Gladys Lacks, Ruby Lacks, Thurl Lacks, Polly Martin, Sadie Sturdivant, John and Dolly Terry, and Peter Wooden. Special thanks to Cliff Garret, a wonderful storyteller who helped bring Henrietta's youth and old Clover to life for me, and always made me smile. Thanks also to Christine Pleasant Tonkin, a distant relative of Henrietta Lacks who traced the Pleasant side of Henrietta's family back to its slave ancestors and generously shared her research with me; she also read the manuscript and provided many valuable suggestions. And to Courtney Speed for her enthusiasm, for sharing her story, and for gathering others to talk with me.

I feel lucky to have found Mary Kubicek, whose sharp memory, tireless patience, and enthusiasm were invaluable. The same is true of George Gey Jr. and his sister, Frances Greene. I'm very fortunate that they spent much of their childhood in the Gey lab with their parents and were able to bring those years to life for me. Thanks also to Frances's husband, Frank Greene.

I'm very grateful to the many librarians and archivists who took the time to track down old newspaper and journal articles, photos, videos, and other resources. Special thanks to Andy Harrison, curator of the George Gey collection at the Alan Mason Chesney Medical Archives; to former University of Pittsburgh library sciences students Amy Notarius and Elaina Vitale; to Frances Woltz, who provided me with a wealth of information and stories; and to Hap Hagood, Phoebe Evans Letocha, and Tim Wisniewski. David Smith at the New York Public Library helped me as he has many other lucky writers, and secured me a quiet workspace in the library's Wertheim Study. David Rose, archivist for the March of Dimes Foundation, took such a deep interest in this book that he conducted hours' worth of helpful research on my behalf. To him I owe tremendous gratitude (and lunch).

Hundreds of people gave generously of their time for interviews, and I thank them all, particularly George Annas, Laure Aurelian, Baruch Blumberg, Ellen Wright Clayton, Nathanial Comfort, Louis Diggs, Bob Gellman, Carol Greider, Michael Grodin, Wayne Grody, Cal Harley, Robert Hay, Kathy Hudson, Grover Hutchins, Richard Kidwell, David Korn, Robert Kurman, John Masters, Stephen O'Brien, Anna O'Connell, Robert Pollack, John Rash, Judith Greenberg, Paul Lurz, Todd Savitt, Terry Sharrer, Mark Sobel, Robert Weir, Barbara Wyche, and Julius Youngner. For their time, encouragement, and expertise I give special thanks to Lori Andrews, Ruth Faden, and Lisa Parker, who spurred my thinking with early conversations, and read the manuscript, offering helpful comments. Thanks also to Duncan Wilson, who provided me with an early version of his dissertation and some very helpful research materials.

Several scientists deserve special thanks: Howard W. Jones, Victor McKusick, and Susan Hsu shared invaluable memories; all were unflinchingly honest and patient with my many questions. Leonard Hayflick spent more than a dozen hours on the phone with me, often taking my calls when he was traveling or in the midst of his own work. His memory and scientific expertise were a tremendous resource. He offered extremely valuable comments on a draft of this book, as did Robert Stevenson, who supported this project from the beginning, when not all scientists did. He was an enormous asset.

I'm grateful to Roland Pattillo for taking the time to figure me out, for believing in me, for schooling me, and for helping me contact Deborah. He and his wife, Pat, opened themselves and their home to me early on, and have been supportive since. They also read a draft of the book and offered helpful suggestions.

Christoph Lengauer's passion and his willingness to be swept into the Lackses' story were inspiring. I thank him for his patience, openness, and forward thinking. He answered many questions and read this book in draft form, offering honest and extremely helpful feedback.

Several writers who have covered the HeLa story were generous with their time. Michael Gold wrote about the contamination story in great detail in his book, *A Conspiracy of Cells,* which was a wonderful resource. It was always a joy to talk with Michael Rogers, whose 1976 *Rolling Stone* article about HeLa was an important resource when I began working on this book. Harriet Washington, author of *Medical Apartheid,* has been a wonderful champion of this book; she talked with me about her experience interviewing the Lacks family for a 1994 *Emerge* article, and offered helpful comments on a draft of the book.

Special thanks to Ethan Skerry and Lowenstein Sandler PC for the pro bono work they did to help me establish the Henrietta Lacks Foundation. Thanks to the University of Memphis for a grant that helped with final research and fact-checking for this book. I'm grateful to both my students and colleagues, particularly Kristen Iversen and Richard Bausch, wonderful teachers, writers, and friends. Special thanks to John Calderazzo and Lee Gutkind for more than a decade of encouragement, support, and close friendship. John realized I was a writer long before I did, and has always been an inspiration. Lee taught me to care deeply about story structure and gave me entrée into the worlds of professional writing, and working at 5:00 A.M. Many thanks also to Donald Defler, for introducing me to Henrietta, and teaching biology with passion.

This book was intensively fact-checked. As part of that process, many experts read it before publication to help ensure its accuracy. I thank them for their time and valuable feedback: Erik Angner (a close friend and strong supporter of this book from its inception), Stanley Gartler, Linda MacDonald Glenn, Jerry Menikoff, Linda Griffith, Miriam Kelty (who also provided helpful documents from her personal archive), Joanne Manaster (aka @sciencegoddess), Alondra Nelson (who deserves special thanks for her honesty, and for saving me from a serious omission), Rich Purcell, Omar Quintero (who also provided beautiful HeLa photos and video footage for the book and its website), Laura Stark, and Keith Woods. Thanks also to the many people who read selected chapters, particularly Nathaniel Comfort and Hannah Landecker (whose extensive work on HeLa and the history of cell culture, especially her book, *Culturing Life,* was a tremendous resource).

Every writer should be lucky enough to find an expert source as generous with his time as Vincent Racaniello. He read multiple drafts, sent many resources, and offered invaluable feedback. His belief in the importance of communicating science to the general public in an accurate and accessible way (witnessed in his "This Week in Virology" podcasts at TWiV.tv and his Twitter feed @profvrr) is a great model for other scientists. The same is true for David Kroll (@abelpharmboy), a big supporter of this book, who writes about science on his blog, Scienceblogs.com/terrasig. He provided helpful feedback and research material, and even took his scanner to a library to gather a few key documents for me. I feel very fortunate to call him a friend.

My graduate assistant Leigh Ann Vanscoy dove into her job with great enthusiasm, working hard to track down photos and permissions, and helping with fact-checking during the final hours. Pat Walters (patwalters.net), research assistant extraordinaire, talented young writer and reporter, and good friend, fact-checked this entire book and devoted himself to the process with unparalleled enthusiasm, precision, and attention to detail. He dug out hard-to-find facts, and his work saved me from numerous errors (including my apparent inability to do basic math). This book benefited greatly from his contributions. I'm lucky to have found him, and I look forward to seeing his bright future unfold.

Several other people helped with research and fact-checking and I thank them all. The great Charles Wilson at *The New York Times Magazine* fact-checked the portions of this book that originally appeared in the magazine, and was a joy to work with. Heather Harris acted as my stand-in when I couldn't get to Baltimore, doggedly gathering court and archival documents, often on short notice. Av Brown of yourmaninthestacks.com was, indeed, my man in the stacks, always thorough and fast with research requests. Paige Williams swooped in to help with some last-minute fact-checking in the midst of her own busy writing career. And my longtime friend Lisa Thorne deserves special thanks (and probably some wrist splints) for transcribing the majority of my interview tapes and offering wonderful commentary on what she heard.

I'm thankful to many great reporters, writers, and editors who offered encouragement, advice, feedback, and friendship along the way, particularly Jad Abumrad, Alan Burdick, Lisa Davis, Nicole Dyer, Jenny Everett, Jonathan Franzen, Elizabeth Gilbert, Cindy Gill, Andrew

Hearst, Don Hoyt Gorman, Alison Gwinn, Robert Krulwich, Robin Marantz Henig, Mark Jannot, Albert Lee, Erica Lloyd, Joyce Maynard, James McBride, Robin Michaelson, Gregory Mone, Michael Moyer, Scott Mowbray, Katie Orenstein, Adam Penenberg, Michael Pollan, Corey Powell, Mark Rotella, Lizzie Skurnick, Stacy Sullivan, Paul Tough, Jonathan Weiner, and Barry Yeoman. Special thanks to Dinty W. Moore, Diana Hume George, and the many other wonderful writers I taught with at the now-sadly-defunct Mid-Atlantic Creative Nonfiction Summer Writers Conference. I miss you all. Thanks also to the editors who worked with me on my early stories related to the book: Patti Cohen at the *New York Times,* Sue De Pasquale at *Johns Hopkins Magazine,* Sally Flecker at *Pitt Magazine,* and James Ryerson at *The New York Times Magazine,* who always makes my work better. Also to my fellow bloggers on ScienceBlogs.com, the ever helpful and inspiring Invisible Institute, the amazing Birders, and my wonderful Facebook and Twitter friends, who provided resources, laughter, encouragement, and celebration of moments big and small. Thanks also to Jon Gluck for helpful early editorial advice. And to Jackie Heinze, who amazingly gave me her car so I could disappear into the middle of nowhere for months to write. Special thanks to Albert French, who helped me take the first difficult steps toward writing this book by challenging me to a race and letting me win.

I owe deep gratitude to all of my former colleagues on the National Book Critics Circle board of directors, whose devotion to great books helped keep me inspired, motivated, and thinking critically. Special thanks to Rebecca Miller, Marcela Valdes, and Art Winslow, who provided years of encouragement, read drafts of the book, and offered insightful comments. As did John Freeman, who I thank for the hours we spent talking about writing and this book, and for Ford and friendship.

My agent, Simon Lipskar at Writers House, has my endless thanks for fighting with and for me when others wouldn't, for being a rock star and a friend. I knew there was a reason I liked you. As is true for many books these days, mine struggled to find its way to press. Three publishing houses and four editors later, I feel extremely lucky to have landed at Crown with Rachel Klayman as my editor. She inherited my book, immediately adopted it as her own, and never faltered in her support of it. She devoted more of her time and heart to this book than I could have dreamed of. Every writer should be fortunate enough to work with such a talented editor, and to have a publishing house as devoted as Crown has been. I'm deeply grateful to everyone on Team Immortal at Crown: their passion for this book and the incredible work they did to send it into the world as best they could has been astonishing and humbling. Special thanks to Tina Constable for her undying support, and for being there for the long haul; to Courtney Greenhalgh, my wonderful and tireless publicist; to Patty Berg, for her creative pursuit of every marketing opportunity; and to Amy Boorstein, Jacob Bronstein, Stephanie Chan, Whitney Cookman, Jill Flaxman, Matthew Martin, Philip Patrick, Annsley Rosner, Courtney Snyder, Barbara Sturman, Katie Wainwright, and Ada Yonenaka. I feel so fortunate to have worked with you all. The same is true of Leila Lee and Michael Gentile in the academic marketing department at Random House, who believed in this book and worked hard to help get it into classrooms. Thanks also to the Random House sales force, particularly John Hastie, Michael Kindness, Gianna LaMorte, and Michele Sulka, who embraced this book and ran with it.

I'm deeply grateful to Erika Goldman, Jon Michel, and Bob Podrasky, all formerly at W. H. Freeman, for believing in me and this book from the beginning, and encouraging me to fight for what I wanted it to be. Thanks also to Louise Quayle for her help early in the process, and to Caroline Sincerbeaux, for always loving this book, and for bringing it to Crown where it found a wonderful home.

Betsy and Michael Hurley and the Lancaster Literary Guild deserve far more thanks than I could possibly convey here. They gave me a key to writer heaven: a beautiful retreat in the hills of West Virginia, where I was free to write without distraction, often for months on end. The world would be a better place if more organizations like the Lancaster Literary Guild existed to support the arts. Along with that retreat house came amazing neighbors: Joe and Lou Rable kept me safe, full, happy, and loved. Jeff and Jill Shade helped me stay human during months of endless work, providing friendship and fun, beautiful property to walk my dogs on, and Baristas and JJS Massage, my favorite café in the world, where Jill kept me well fed and caffeinated, and Jeff massaged the knots he called "writers' blocks" from my arms, poured drinks when I needed them, and talked with me for hours about my book. I thank the town of New Martinsville, West Virginia, for taking me in. And Heather at The Book Store, who tracked down every good novel she could find with a disjointed structure, all of which I devoured while trying to figure out the structure of this book.

I am lucky to have many wonderful friends who were tireless cheerleaders for this project, despite the number of times they heard me say, "I can't, because I have to work on my book." I thank them all, particularly Anna Bargagliotti, Zvi Biener, Stiven Foster (Celebration Committee!), Ondine Geary, Peter Machamer, Jessica Mesman (Foo!), Jeff and Linda Miller, Elise Mittleman (P and PO!), Irina Reyn, Heather Nolan (who also read an early draft and offered helpful feedback), Andrea Scarantino, Elissa Thorndike, and John Zibell. I'm grateful to Gualtiero Piccinini for encouragement and support early in the book process. Special thanks to my dear friend Stephanie Kleeschulte, who brings me joy and keeps me young. And to Quail Rogers-Bloch, for our history, for laughter, wine, and stupid movies in the midst of madness (Yes he did, sir!). Without her, I wouldn't be who I am today. She gave me a home to return to each night after my work in Baltimore, talked me through the hardest parts of this book, rescued me when I got stranded or ran out of money, and always offered wise feedback on drafts (some of which she listened to over the phone). Her wonderful husband, Gyon, fed me mangoes when I was exhausted, and their son, my godson, Aryo, brought much joy. Quail's mother, Terry Rogers, always an inspiration, also provided wonderful feedback on this book.

I'm very lucky to count Mike Rosenwald (mikerosenwald.com) as one of my closest friends. He's an inspiration as a writer, reporter, and reader. He's been with me every step of this book with encouragement, commiseration, advice, and a few much-needed ass-kickings. He read many drafts (and listened to several sections over the phone), always offering helpful feedback. I look forward to returning the favor.

My family was the backbone of this book: Matt, the best big brother a girl could hope for, supported me with long talks and laughter and always reminded me to watch out for myself. My wonderful nephews, Nick and Justin, never fail to bring me joy. They spent far too many holidays without their aunt because of this book, and I look forward to making up for lost time. My sister-in-law, Renée, has provided never-ending support for this book; she is not only a good friend, but an eagle-eyed reader with an incredible talent for spotting errors and inconsistencies. The same is true of my wonderful stepmother, Beverly, who read several drafts, giving invaluable support and insight. I also benefited greatly from her sensitivity and training as a social worker as I navigated the complexities of the Lackses' experience.

My parents and their spouses deserve to have entire wings of this book named after them for all the support they've given me over the years. My mother, Betsy McCarthy, has never faltered in her belief in me and this book. She's kept me sane through pep talks, reality checks, and the gift of knitting, a family tradition I treasure. Her drive, her artistry, and her determination have been a tremendous guide for me. She and her husband, Terry, encouraged me during the hardest times, read multiple drafts of the book, and provided wise and helpful feedback.

I am endlessly thankful to my father, Floyd Skloot, for teaching me to see the world with a writer's eyes, for inspiring me with his many wonderful books, and for treating this one as if it were his own. He has always encouraged me to follow my art, and to fight for what I believed it could be, even when that meant taking risks like quitting a stable job to freelance. He read this book six times before publication (and that's not counting dozens of individual chapters and sections he read before that). He is not only my father but my colleague, my selfless publicist, and my friend. For that I am lucky beyond measure.

And then there's David Prete, my Focus (*you* know). He read this manuscript when it was far longer than any book should ever be, and used his rich talent as a writer and an actor to help me get it to a manageable size. With his grace and support, his heart, his compassion, and his amazing cooking, he also kept me alive and happy. Even when The Immortal Book Project of Rebecca Skloot took over our home and lives, his support never wavered. He has my love and my gratitude. I am a very lucky woman.

Notes

The source materials I relied on to write this book filled multiple file cabinets, and the hundreds of hours of interviews I conducted—with members of the Lacks family, scientists, journalists, legal scholars, bioethicists, health policy experts, and historians—fill several shelves worth of notebooks. I have not listed all of those experts in these notes, but many are thanked in the acknowledgments or cited by name in the book.

Because my sources are too extensive to list in their entirety, these notes feature a selection of some of the most valuable, with a focus on those that are publicly available. For additional information and resources, visit RebeccaSkloot.com.

These notes are organized by chapter, with two exceptions: Since the Lacks family and George Gey appear throughout many chapters, I have consolidated my notes about them and listed them immediately below. If a chapter is not listed in the notes, it means the source material for that chapter is described in these consolidated entries about Gey and the Lackses.

Henrietta Lacks and Her Family
To re-create the story of Henrietta's life and the lives of her relatives, I relied on interviews with her family, friends, neighbors, and experts on the time and place in which they lived, as well as family audio and video recordings, and unedited B-roll from the BBC documentary *The Way of All Flesh*. I also relied on the journals of Deborah Lacks, medical records, court documents, police records, family photographs, newspaper and magazine reports, community newsletters, wills, deeds, and birth and death certificates.

George Gey and His Lab
To re-create the lives and work of George and Margaret Gey, I relied on the holdings of the George Gey archives at the Alan Mason Chesney Medical Archives (AMCMA) at Johns Hopkins Medical School; the Tissue Culture Association Archives (TCAA) at the University of Maryland, Baltimore County; the personal archives of Gey's family; as well as on academic papers, and interviews with family, colleagues, and scientists in the fields of cancer research and cell culture.

Prologue
The estimate of the possible weight of HeLa cells comes from Leonard Hayflick, who calculated the greatest possible weight potential of a normal human cell strain as 20 million metric tons and says HeLa's potential would be "infinitely greater" since it's not bound by the Hayflick limit. As Hayflick wrote to me in an email: "If we were to grow HeLa for just 50 population doublings it would yield 50 million metric tons if all the cells were saved. Clearly that is impractical to do." For more information on the growth potential of a normal cell, see Hayflick and Moorehead, "The Serial Cultivation of Human Diploid Cell Strains," *Experimental Cell Research* 25 (1961).

For the articles about the Lacks family I refer to, see "Miracle of HeLa," *Ebony* (June 1976) and "Family Takes Pride in Mrs. Lacks' Contribution," *Jet* (April 1976).

PART ONE: LIFE

Chapter 1: The Exam
Conflicting dates have been reported for Henrietta's first visit to Johns Hopkins; the date most commonly cited is February 1, 1951. The lack of clarity surrounding the date results from a transcription error noted by her doctor on February 5. Elsewhere her records indicate that her tumor was first tested on January 29, so I have used that date.

For documentation of the history of Johns Hopkins (in this and later chapters), see the AMCMA, as well as *The Johns Hopkins Hospital and the Johns Hopkins University School of Medicine: A Chronicle,* by Alan Mason Chesney, and *The First 100 Years: Department of Gynecology and Obstetrics, the Johns Hopkins School of Medicine, the Johns Hopkins Hospital,* edited by Timothy R. B. Johnson, John A. Rock, and J. Donald Woodruff.

Information here and in later chapters regarding segregation at Johns Hopkins came from interviews as well as from Louise Cavagnaro, "The Way We Were," *Dome* 55, no. 7 (September 2004), available at hopkinsmedicine.org/dome/0409/feature1.cfm; Louise Cavagnaro, "A History of Segregation and Desegregation at The Johns Hopkins Medical Institutions," unpublished manuscript (1989) at the AMCMA; and "The Racial Record of Johns Hopkins University," *Journal of Blacks in Higher Education* 25 (Autumn 1999).

Sources on the effects segregation had on health-care delivery and outcomes include: *The Strange Career of Jim Crow,* by C. Vann Woodward; P. Preston Reynolds and Raymond Bernard, "Consequences of Racial Segregation," *American Catholic Sociological Review* 10, no. 2 (June 1949); Albert W. Dent, "Hospital Services and Facilities Available to Negroes in the United States," *Journal of Negro Education* 18, no. 3 (Summer 1949); Alfred Yankauer Jr., "The Relationship of Fetal and Infant Mortality to Residential Segregation: An Inquiry into Social Epidemiology," *American Sociological Review* 15, no. 5 (October 1950); and "Hospitals and Civil Rights, 1945–1963: The Case of Simkins v. Moses H. Cone Memorial Hospital," *Annals of Internal Medicine* 126, no. 11 (June 1, 1997).

Henrietta's medical records, provided to me by her family, are not publicly available, but some information on her diagnosis can be found in Howard W. Jones, "Record of the First Physician to see Henrietta Lacks at the Johns Hopkins Hospital: History of the Beginning of the HeLa Cell Line," *American Journal of Obstetrics and Gynecology* 176, no. 6 (June 1997): S227–S228.

Chapter 2: Clover

Information on the history of Virginia tobacco production came from the Virginia Historical Society, the Halifax County website, archival documents and news articles at the South Boston Library, and several books, including *Cigarettes: Anatomy of an Industry, from Seed to Smoke,* by Tara Parker Pope, an overview of tobacco history for the general public.

Several books helped me reconstruct the era and places in which Henrietta lived, including *Country Folks: The Way We Were Back Then in Halifax County, Virginia,* by Henry Preston Young, Jr; *History of Halifax,* by Pocahontas Wight Edmunds; *Turner Station,* by Jerome Watson; *Wives of Steel,* by Karen Olson; and *Making Steel,* by Mark Reutter. The history of Turner Station is also chronicled in news articles and documents housed at the Dundalk Patapsco Neck Historical Society and the North Point Library in Dundalk, Maryland.

Chapter 3: Diagnosis and Treatment

For information on the development of the Pap smear, see G. N. Papanicolaou and H. F. Traut, "Diagnostic Value of Vaginal Smears in Carcinoma of Uterus," *American Journal of Obstetrics and Gynecology* 42 (1941), and "Diagnosis of Uterine Cancer by the Vaginal Smear," by George Papanicolaou and H. Traut (1943).

Richard TeLinde's research on carcinoma in situ and invasive carcinoma, and his concern about unnecessary hysterectomies, is documented in many papers, including "Hysterectomy: Present-Day Indications," *JMSMS* (July 1949); G. A. Gavin, H. W. Jones, and R. W. TeLinde, "Clinical Relationship of Carcinoma in Situ and Invasive Carcinoma of the Cervix," *Journal of the American Medical Association* 149, no. 8 (June 2, 1952); R. W. TeLinde,

H. W. Jones and G. A. Gavin, "What Are the Earliest Endometrial Changes to Justify a Diagnosis of Endometrial Cancer?" *American Journal of Obstetrics and Gynecology* 66, no. 5 (November 1953); and TeLinde, "Carcinoma in Situ of the Cervix," *Obstetrics and Gynecology* 1, no. 1 (January 1953); also the biography *Richard Wesley TeLinde,* by Howard W. Jones, Georgeanna Jones, and William E. Ticknor.

For information on the history of radium and its use as a cancer treatment, see *The First 100 Years;* the website of the U. S. Environmental Protection Agency at epa.gov/iris/subst/0295.htm; D. J. DiSantis and D. M. DiSantis, "Radiologic History Exhibit: Wrong Turns on Radiology's Road of Progress," *Radiographics* 11 (1991); and *Multiple Exposures: Chronicles of the Radiation Age,* by Catherine Caufield.

Sources on the standard treatment regimen for cervical cancer in the 1950s include A. Brunschwig, "The Operative Treatment of Carcinoma of the Cervix: Radical Panhysterectomy with Pelvic Lymph Node Excision," *American Journal of Obstetrics and Gynecology* 61, no. 6 (June 1951); R. W. Green, "Carcinoma of the Cervix: Surgical Treatment (A Review)," *Journal of the Maine Medical Association* 42, no. 11 (November 1952); R. T. Schmidt, "Panhysterectomy in the Treatment of Carcinoma of the Uterine Cervix: Evaluation of Results," *JAMA* 146, no. 14 (August 4, 1951); and S. B. Gusberg and J. A. Corscaden, "The Pathology and Treatment of Adenocarcinoma of the Cervix," *Cancer* 4, no. 5 (September 1951).

Growth of the L-cell (the first immortal cell line, grown from a mouse) was documented in W. R. Earle et al., "Production of Malignancy in Vitro. IV. The Mouse Fibroblast Cultures and Changes Seen in Living Cells," *Journal of the NCI* 4 (1943).

For information about Gey's pre-HeLa cell culture work, see G. O. Gey, "Studies on the Cultivation of Human Tissue Outside the Body," *Wisconsin J.J.* 28, no. 11 (1929); G. O. Gey and M. K. Gey, "The Maintenance of Human Normal Cells and Human Tumor Cells in Continuous Culture I. A Preliminary Report," *American Journal of Cancer* 27, no. 45 (May 1936); an overview can be found in G. Gey, F. Bang, and M. Gey, "An Evaluation of Some Comparative Studies on Cultured Strains of Normal and Malignant Cells in Animals and Man," *Texas Reports on Biology and Medicine* (Winter 1954).

Chapter 4: The Birth of HeLa

For information on Gey's development of the roller drum, see "An Improved Technic for Massive Tissue Culture," *American Journal of Cancer* 17

(1933); for his early work filming cells, see G. O. Gey and W. M. Firor, "Phase Contrast Microscopy of Living Cells," *Annals of Surgery* 125 (1946). For the abstract he eventually published documenting the initial growth of the HeLa cell line, see G. O. Gey, W. D. Coffman, and M. T. Kubicek, "Tissue Culture Studies of the Proliferative Capacity of Cervical Carcinoma and Normal Epithelium," *Cancer Research* 12 (1952): 264–65. For a thorough discussion of his work on HeLa and other cultures, see G. O. Gey, "Some Aspects of the Constitution and Behavior of Normal and Malignant Cells Maintained in Continuous Culture," *The Harvey Lecture Series L* (1954–55).

Chapter 5: "Blackness Be Spreadin All Inside"
TeLinde's discussion of the "psychic effects of hysterectomy" can be found in "Hysterectomy: Present-Day Indications," *Journal of the Michigan State Medical Society,* July 1949.

Chapter 6: "Lady's on the Phone"
Papers from the first HeLa symposium were published in "The HeLa Cancer Control Symposium: Presented at the First Annual Women's Health Conference, Morehouse School of Medicine, October 11, 1996," edited by Roland Pattillo, *American Journal of Obstetrics and Gynecology* suppl. 176, no. 6 (June 1997).

For an overview of the Tuskegee study aimed at the general public, see *Bad Blood: The Tuskegee Syphilis Experiment,* by James H. Jones; see also "Final Report of the Tuskegee Syphilis Study Legacy Committee," Vanessa Northington Gamble, chair (May 20, 1996).

Chapter 7: The Death and Life of Cell Culture
For the television segment featuring George Gey, see "Cancer Will Be Conquered," *Johns Hopkins University: Special Collections Science Review Series* (April 10, 1951).

For additional reading on the history of cell culture, see *Culturing Life: How Cells Became Technologies,* by Hannah Landecker, the definitive history; also see *The Immortalists: Charles Lindberg, Dr. Alexis Carrel, and Their Daring Quest to Live Forever,* by David M. Friedman. For a general overview of Hopkins's contributions to cell culture, see "History of Tissue Culture at Johns Hopkins," *Bulletin of the History of Medicine* (1977).

To re-create the story of Alexis Carrel and his chicken heart, I relied on these sources and many others: A. Carrel and M. T. Burrows, "Cultivation of Tissues in Vitro and Its Technique," *Journal of Experimental Medicine* (January 15, 1911); "On the Permanent Life of Tissues Outside of the Organism," *Journal of Experimental Medicine* (March 15, 1912); Albert H. Ebeling, "A Ten Year Old Strain of Fibroblasts," *Journal of Experimental Medicine* (May 30, 1922), and "Dr. Carrel's Immortal Chicken Heart," *Scientific American* (January 1942); "The 'Immortality' of Tissues," *Scientific American* (October 26, 1912); "On the Trail of Immortality," *McClure's* (January 1913); "Herald of Immortality Foresees Suspended Animation," *Newsweek* (December 21, 1935); "Flesh That Is Immortal," *World's Work* 28 (October 1914); "Carrel's New Miracle Points Way to Avert Old Age!" *New York Times Magazine* (September 14, 1913); Alexis Carrel, "The Immortality of Animal Tissue, and Its Significance," *The Golden Book Magazine* 7 (June 1928); and "Men in Black," *Time* 31, number 24 (June 13, 1938). The Nobel Prize website also contains much useful information about Carrel.

For a history of cell culture in Europe, see W. Duncan, "The Early History of Tissue Culture in Britain: The Interwar Years," *Social History of Medicine* 18, no. 2 (2005), and Duncan Wilson, " 'Make Dry Bones Live': Scientists' Responses to Changing Cultural Representation of Tissue Culture in Britain, 1918–2004," dissertation, University of Manchester (2005).

The conclusion that Carrel's chicken-heart cells were not actually immortal comes from interviews with Leonard Hayflick; also J. Witkowski, "The Myth of Cell Immortality," *Trends in Biochemical Sciences* (July 1985), and J. Witkowski, letter to the editor, *Science* 247 (March 23, 1990).

Chapter 9: Turner Station
The newspaper article that documented Henrietta's address was Jacques Kelly, "Her Cells Made Her Immortal," *Baltimore Sun*, March 18, 1997. The article by Michael Rogers was "The Double-Edged Helix," *Rolling Stone* (March 25, 1976).

Chapter 10: The Other Side of the Tracks
For reports of the decline of Clover, see, for example, "South Boston, Halifax County, Virginia," an Economic Study by Virginia Electric and Power Company; "Town Begins to Move Ahead," *Gazette-Virginian* (May 23, 1974); "Town Wants to Disappear," *Washington Times* (May 15, 1988); and "Supes Decision Could End Clover's Township," *Gazette-Virginian* (May 18, 1998); "Historical Monograph: Black Walnut Plantation Rural Historic District, Halifax County, Virginia," Old Dominion Electric Cooperative (April 1996). Population figures are available at census.gov.

PART TWO: DEATH

Chapter 12: The Storm

For a discussion of the history of court decisions and rights regarding autopsies, see *Subjected to Science,* by Susan Lederer.

Chapter 13: The HeLa Factory

For further reading on the history of the polio vaccine, see *The Virus and the Vaccine,* by Debbie Bookchin and Jim Shumacher; *Polio: An American Story,* by David M. Oshinski; *Splendid Solution: Jonas Salk and the Conquest of Polio,* by Jeffrey Kluger; and *The Cutter Incident: How America's First Polio Vaccine Led to the Growing Crisis in Vaccines,* by Paul Offit.

Details of the initial growth of poliovirus using HeLa cells, and the subsequent development of shipping methods, is documented in letters housed at the AMCMA and the March of Dimes Archives, as well as in J. Syverton, W. Scherer, and G. O. Gey, "Studies on the Propagation in Vitro of Poliomyelitis Virus," *Journal of Experimental Medicine* 97, no. 5 (May 1, 1953).

The history of the HeLa mass production facilities at Tuskegee is documented in letters, expense reports, and other documents at the March of Dimes Archives. For a comprehensive overview, see Russell W. Brown and James H. M. Henderson, "The Mass Production and Distribution of HeLa Cells at the Tuskegee Institute, 1953–55," *Journal of the History of Medicine* 38 (1983).

A detailed history of many scientific advances that followed the growth of HeLa can be found in letters and other papers in the AMCA and TCAA. The book *Culturing Life: How Cells Became Technologies,* by Hannah Landecker, provides a comprehensive overview. Many of the scientific papers referred to in this chapter are collected in *Readings in Mammalian Cell Culture,* edited by Robert Pollack, including H. Eagle, "Nutrition Needs of Mammalian Cells in Tissue Culture," *Science* 122 (1955): 501–4; T. T. Puck and P. I. Marcus, "A Rapid Method for Viable Cell Titration and Clone Production with HeLa Cells in Tissue Culture: The Use of X-irradiated Cells to Study Conditioning Factors," *Proceedings of the National Academy of Science* 41 (1955); J. H. Tjio and A. Levan, "The Chromosome Number of Man," *Cytogenics* 42 (January 26, 1956). See also M. J. Kottler, "From 48 to 46: Cytological Technique, Preconception, and the Counting of Human Chromosomes," *Bulletin of the History of Medicine* 48, no. 4 (1974); H. E. Swim, "Microbiological Aspects of Tissue Culture," *Annual Review of Microbiology* 13 (1959); J. Craigie, "Survival and Preservation of Tumors in the Frozen State," *Advanced Cancer Research* 2 (1954); W. Scherer and

A. Hoogasian, "Preservation at Subzero Temperatures of Mouse Fibroblasts (Strain L) and Human Epithelial Cells (Strain HeLa)," *Proceedings of the Society for Experimental Biology and Medicine* 87, no. 2 (*1954*); T. C. Hsu, "Mammalian Chromosomes in Vitro: The Karyotype of Man," *Journal of Heredity* 43 (1952); and D. Pearlman, "Value of Mammalian Cell Culture as Biochemical Tool," *Science* 160 (April 1969); and N. P. Salzman, "Animal Cell Cultures," *Science* 133, no. 3464 (May 1961).

Other useful resources for this chapter include *Human and Mammalian Cytogenetics: An Historical Perspective,* by T. C. Hsu; and C. Moberg, "Keith Porter and the Founding of the Tissue Culture Association: A Fiftieth Anniversary Tribute, 1946–1996," *In Vitro Cellular & Developmental Biology–Animal* (November 1996).

Chapter 14: Helen Lane
The debate about releasing Henrietta's name to the public is documented in letters located in the AMCA. The article that identified "Henrietta Lakes" as the source of the HeLa cell line was "U Polio-detection Method to Aid in Prevention Plans," *Minneapolis Star,* November 2, 1953. The first article to identify "Helen L." as the source of the HeLa cell line was Bill Davidson, "Probing the Secret of Life," *Collier's,* May 14, 1954.

Chapter 17: Illegal, Immoral, and Deplorable
Southam's cancer cell injections are documented in many scientific articles he authored or coauthored, including "Neoplastic Changes Developing in Epithelial Cell Lines Derived from Normal Persons," *Science* 124, no. 3212 (July 20, 1956); "Transplantation of Human Tumors," letter, *Science* 125, no. 3239 (January 25, 1957); "Homotransplantation of Human Cell Lines," *Science* 125, no. 3239 (January 25, 1957); "Applications of Immunology to Clinical Cancer Past Attempts and Future Possibilities," *Cancer Research* 21 (October 1961): 1302–16; and "History and Prospects of Immunotherapy of Cancer," *Annals of the New York Academy of Sciences* 277, no. 1 (1976).

For media coverage of Southam's prison studies, see "Convicts to Get Cancer Injection," *New York Times,* May 23, 1956; "Cancer by the Needle," *Newsweek,* June 4, 1956; "14 Convicts Injected with Live Cancer Cells," *New York Times,* June 15, 1956; "Cancer Volunteers," *Time,* February 25, 1957; "Cancer Defenses Found to Differ," *New York Times,* April 15, 1957; "Cancer Injections Cause 'Reaction,'" *New York Times,* July 18, 1956; "Convicts Sought for Cancer Test," *New York Times,* August 1, 1957.

The most complete resource on Southam's cancer cell injections and the hearings that followed is *Experimentation with Human Beings,* by Jay

Katz, in which he collected extensive original correspondence, court documents, and other materials that might otherwise have been lost, as they weren't retained by the Board of Regents. Also see Jay Katz, "Experimentation on Human Beings," *Stanford Law Review* 20 (November 1967). For Hyman's lawsuits, see *William A. Hyman v. Jewish Chronic Disease Hospital* (42 Misc. 2d 427; 248 N.Y.S.2d 245; 1964 and 15 N.Y.2d 317; 206 N.E.2d 338; 258 N.Y.S.2d 397; 1965). Also see patient lawsuit, *Alvin Zeleznik v. Jewish Chronic Disease Hospital* (47 A.D.2d 199; 366 N.Y.S.2d 163; 1975). Beecher's paper is H. Beecher, "Ethics and Clinical Research," *New England Journal of Medicine* 274, no. 24 (June 16, 1966).

The news coverage of the ethical debate surrounding the Southam controversy includes "Scientific Experts Condemn Ethics of Cancer Injection," *New York Times,* January 26, 1964; Earl Ubell, "Why the Big Fuss," *Chronicle-Telegram,* January 25, 1961; Elinor Langer, "Human Experimentation: Cancer Studies at Sloan-Kettering Stir Public Debate on Medical Ethics," *Science* 143 (February 7, 1964); and Elinor Langer, "Human Experimentation: New York Verdict Affirms Patient Rights," *Science* (February 11, 1966).

Susan E. Lederer's *Subjected to Science: Human Experimentation in America Before the Second World War* is a must-read on the ethics and history of research on human subjects, as is George J. Annas and Michael A. Grodin's *The Nazi Doctors and the Nuremberg Code: Human Rights in Human Experimentation.* Both were important sources for this chapter. For the history of experimentation on prisoners, see *Acres of Skin: Human Experiments at Holmesburg Prison,* by Allen Hornblum, who interviewed Southam before he died, and kindly shared information from those interviews with me.

For further reading in the history of bioethics, including the changes that followed the Southam controversy, see Albert R. Jonsen's *The Birth of Bioethics*; David J. Rothman's *Strangers at the Bedside: A History of How Law and Bioethics Transformed Medical Decision Making*; George J. Annas's *Informed Consent to Human Experimentation: The Subject's Dilemma*; M. S. Frankel, "The Development of Policy Guidelines Governing Human Experimentation in the United States: A Case Study of Public Policy-making for Science and Technology," *Ethics in Science and Medicine* 2, no. 48 (1975); and R. B. Livingston, "Progress Report on Survey of Moral and Ethical Aspects of Clinical Investigation: Memorandum to Director, NIH" (November 4, 1964).

For the definitive history of informed consent, see Ruth Faden and Tom Beauchamp's *A History and Theory of Informed Consent.* For the first court case mentioning "informed consent," see *Salgo v. Leland Stanford Jr. University Board of Trustees* (Civ. No. 17045. First Dist., Div. One, 1957).

Chapter 18: The Strangest Hybrid

Instructions for growing HeLa at home were published in C. L. Stong, "The Amateur Scientist: How to Perform Experiments with Animal Cells Living in Tissue Culture," *Scientific American,* April 1966.

Sources documenting the history of cell culture research in space include Allan A. Katzberg, "The Effects of Space Flights on Living Human Cells," Lectures in Aerospace Medicine, School of Aerospace Medicine (1960); and K. Dickson, "Summary of Biological Spaceflight Experiments with Cells," *ASGSB Bulletin* 4, no. 2 (July 1991).

Though the research done on HeLa cells in space was legitimate and useful, we now know that it was part of a cover-up for a reconnaissance project that involved photographing the Soviet Union from space. For information on the use of "biological payloads" as cover for spy missions, see *Eye in the Sky: The Story of the Corona Spy Satellites,* edited by Dwayne A. Day et al.

The early paper suggesting the possibility of HeLa contamination is L. Coriell et al., "Common Antigens in Tissue Culture Cell Lines," *Science,* July 25, 1958. Other resources related to early concern over culture contamination include L. B. Robinson et al., "Contamination of Human Cell Cultures by Pleuropneumonialike Organisms," *Science* 124, no. 3232 (December 7, 1956); R. R. Gurner, R. A. Coombs, and R. Stevenson, "Results of Tests for the Species of Origins of Cell Lines by Means of the Mixed Agglutination Reaction," *Experimental Cell Research* 28 (September 1962); R. Dulbecco, "Transformation of Cells in Vitro by Viruses," *Science* 142 (November 15, 1963); R. Stevenson, "Cell Culture Collection Committee in the United States," in *Cancer Cells in Culture,* edited by H. Katsuta (1968). For the history of the ATCC, see R. Stevenson, "Collection, Preservation, Characterization and Distribution of Cell Cultures," *Proceedings, Symposium on the Characterization and Uses of Human Diploid Cell Strains: Opatija* (1963); and W. Clark and D. Geary, "The Story of the American Type Culture Collection: Its History and Development (1899–1973)," *Advances in Applied Microbiology* 17 (1974).

Important sources on early cell hybrid research include Barski, Sorieul, and Cornefert, "Production of Cells of a 'Hybrid' Nature in Cultures in Vitro of 2 Cellular Strains in Combination," *Comptes Rendus Hebdomadaires des Séances de l'Académie des Sciences* 215 (October 24, 1960); H. Harris and J. F. Watkins, "Hybrid Cells Derived from Mouse and Man: Artificial

Heterokaryons of Mammalian Cells from Different Species," *Nature* 205 (February 13, 1965); M. Weiss and H. Green, "Human-Mouse Hybrid Cell Lines Containing Partial Complements of Human Chromosomes and Functioning Human Genes," *Proceedings of the National Academy of Sciences* 58, no. 3 (September 15, 1967); and B. Ephrussi and C. Weiss, "Hybrid Somatic Cells," *Scientific American* 20, no. 4 (April 1969).

For additional information on Harris's hybrid research, see his "The Formation and Characteristics of Hybrid Cells," in *Cell Fusion: The Dunham Lectures (1970); The Cells of the Body: A History of Somatic Cell Genetics;* "Behaviour of Differentiated Nuclei in Heterokaryons of Animal Cells from Different Species," *Nature* 206 (1965); "The Reactivation of the Red Cell Nucleus," *Journal of Cell Science* 2 (1967); and H. Harris and P. R. Harris, "Synthesis of an Enzyme Determined by an Erythrocyte Nucleus in a Hybrid Cell," *Journal of Cell Science* 5 (1966).

Extensive media coverage included "Man-Animal Cells Are Bred in Lab," *The* [London] *Sunday Times* (February 14, 1965); and "Of Mice and Men," *Washington Post* (March 1, 1965).

Chapter 20: The HeLa Bomb

For this chapter I relied on communications and other documents housed at the AMCA and the TCAA, and on "The Proceedings of the Second Decennial Review Conference on Cell Tissue and Organ Culture, The Tissue Culture Association, Held on September 11–15, 1966," *National Cancer Institute Monographs* 58, no. 26 (November 15, 1967).

The vast number of scientific papers about culture contamination include S. M. Gartler, "Apparent HeLa Cell Contamination of Human Heteroploid Cell Lines," *Nature* 217 (February 4, 1968); N. Auersperg and S. M. Gartler, "Isoenzyme Stability in Human Heteroploid Cell Lines," *Experimental Cell Research* 61 (August 1970); E. E. Fraley, S. Ecker, and M. M. Vincent, "Spontaneous in Vitro Neoplastic Transformation of Adult Human Prostatic Epithelium," *Science* 170, no. 3957 (October 30, 1970); A. Yoshida, S. Watanabe, and S. M. Gartler, "Identification of HeLa Cell Glucose 6-phosphate Dehydrogenase," *Biochemical Genetics* 5 (1971); W. D. Peterson et al., "Glucose-6-Phosphate Dehydrogenase Isoenzymes in Human Cell Cultures Determined by Sucrose-Agar Gel and Cellulose Acetate Zymograms," *Proceedings of the Society for Experimental Biology and Medicine* 128, no. 3 (July 1968); Y. Matsuya and H. Green, "Somatic Cell Hybrid Between the Established Human Line D98 (presumptive HeLa) and 3T3," *Science* 163, no. 3868 (February 14, 1969); and C. S. Stulberg, L. Coriell, et al., "The Animal Cell Culture Collection," *In Vitro* 5 (1970).

For a detailed account of the contamination controversy, see *A Conspiracy of Cells*, by Michael Gold.

Chapter 21: Night Doctors

Sources for information about night doctors and the history of black Americans and medical research include *Night Riders in Black Folk History*, by Gladys-Marie Fry; T. L. Savitt, "The Use of Blacks for Medical Experimentation and Demonstration in the Old South," *Journal of Southern History* 48, no. 3 (August 1982); *Medicine and Slavery: The Disease and Health Care of Blacks in Antebellum Virginia;* F. C. Waite, "Grave Robbing in New England," *Medical Library Association Bulletin* (1945); W. M. Cobb, "Surgery and the Negro Physician: Some Parallels in Background," *Journal of the National Medical Association* (May 1951); V. N. Gamble, "A Legacy of Distrust: African Americans and Medical Research," *American Journal of Preventive Medicine* 9 (1993); V. N. Gamble, "Under the Shadow of Tuskegee: African Americans and Health Care," *American Journal of Public Health* 87, no. 11 (November 1997).

For the most detailed and accessible account available, see Harriet Washington's *Medical Apartheid: The Dark History of Medical Experimentation on Black Americans from Colonial Times to the Present*.

For the history of Hopkins, see notes for chapter 1.

For documents and other materials relating to the 1969 ACLU lawsuit over Hopkins's research into a genetic predisposition to criminal activity, see Jay Katz's *Experimentation with Human Beings*, chapter titled "Johns Hopkins University School of Medicine: A Chronicle. Story of Criminal Gene Research." For further reading, see Harriet Washington, "Born for Evil?" in Roelcke and Maio, *Twentieth Century Ethics of Human Subjects Research* (2004).

Sources for the Hopkins lead-study story include court documents and Health and Human Services records, as well as an interview with a source connected to the case, *Ericka Grimes v. Kennedy Kreiger Institute, Inc.* (24-C-99-925 and 24-C-95-66067/CL 193461). See also L. M. Kopelman, "Children as Research Subjects: Moral Disputes, Regulatory Guidance and Recent Court Decisions," *Mount Sinai Medical Journal* (May 2006); and J. Pollak, "The Lead-Based Paint Abatement Repair & Maintenance Study in Baltimore: Historic Framework and Study Design," *Journal of Health Care Law and Policy* (2002).

Chapter 22: "The Fame She So Richly Deserves"

For the paper in which Henrietta's real name was first published, see H. W. Jones, V. A. McKusick, P. S. Harper, and K. D. Wuu, "George Otto Gey (1899–1970): The HeLa Cell and a Reappraisal of Its Origin," *Obstetrics and Gynecology* 38, no. 6 (December 1971). Also see J. Douglas, "Who Was HeLa?" *Nature* 242 (March 9, 1973); and J. Douglas, "HeLa," *Nature* 242 (April 20, 1973), and B. J. C., "HeLa (for Henrietta Lacks)," *Science* 184, no. 4143 (June 21, 1974).

Information regarding the misdiagnosis of Henrietta's cancer and whether that affected her treatment comes from interviews with Howard W. Jones, Roland Pattillo, Robert Kurman, David Fishman, Carmel Cohen, and others. I also relied on several scientific papers, including S. B. Gusberg and J. A. Corscaden, "The Pathology and Treatment of Adenocarcinoma of the Cervix," *Cancer* 4, no. 5 (September 1951).

For sources regarding the HeLa contamination controversy, see notes for chapter 20. The text of the 1971 National Cancer Act can be found at cancer.gov/aboutnci/national-cancer-act-1971/allpages.

Sources regarding the ongoing controversy include L. Coriell, "Cell Repository," *Science* 180, no. 4084 (April 27, 1973); W. A. Nelson-Rees et al., "Banded Marker Chromosomes as Indicators of Intraspecies Cellular Contamination," *Science* 184, no. 4141 (June 7, 1974); K. S. Lavappa et al., "Examination of ATCC Stocks for HeLa Marker Chromosomes in Human Cell Lines," *Nature* 259 (January 22, 1976); W. K. Heneen, "HeLa Cells and Their Possible Contamination of Other Cell Lines: Karyotype Studies," *Hereditas* 82 (1976); W. A. Nelson-Rees and R. R. Flandermeyer, "HeLa Cultures Defined," *Science* 191, no. 4222 (January 9, 1976); M. M. Webber, "Present Status of MA-160 Cell Line: Prostatic Epithelium or HeLa Cells?" *Investigative Urology* 14, no. 5 (March 1977); and W. A. Nelson-Rees, "The Identification and Monitoring of Cell Line Specificity," in *Origin and Natural History of Cell Lines* (Alan R. Liss, Inc., 1978).

I also relied on both published and unpublished reflections by those directly involved in the controversy. Published articles include W. A. Nelson-Rees, "Responsibility for Truth in Research," *Philosophical Transactions of the Royal Society* 356, no. 1410 (June 29, 2001); S. J. O'Brien, "Cell Culture Forensics," *Proceedings of the National Academy of Sciences* 98, no. 14 (July 3, 2001); and R. Chatterjee, "Cell Biology: A Lonely Crusade," *Science* 16, no. 315 (February 16, 2007).

PART THREE: IMMORTALITY

Chapter 23: "It's Alive"
This chapter relied in part on letters housed at the AMCMA, on Deborah Lacks's medical records, and on "Proceedings for the New Haven Conference (1973): First International Workshop on Human Gene Mapping," *Cytogenetics and Cell Genetics* 13 (1974): 1–216.

For information on Victor McKusick's career, see the National Library of Medicine at nlm.nih.gov/news/victor_mckusick_profiles09.html. His genetic database, now called OMIM, can be found at ncbi.nlm.nih.gov/omim/.

For selected documentation of the relevant regulations protecting human subjects in research, see "The Institutional Guide to DHEW Policy on Protection of Human Subjects," DHEW Publication No. (NIH) 72-102 (December 1, 1971); "NIH Guide for Grants and Contracts," U.S. Department of Health, Education, and Welfare, no. 18 (April 14, 1972); "Policies for Protecting All Human Subjects in Research Announced," *NIH Record* (October 9, 1973); and "Protection of Human Subjects," Department of Health, Education, and Welfare, *Federal Register* 39, no. 105, part 2 (May 30, 1974).

For more information on the history of oversight of research on human subjects, see *The Human Radiation Experiments: Final Report of the President's Advisory Committee* (Oxford University Press, available at hss.energy.gov/HealthSafety/ohre/roadmap/index.html).

Chapter 24: "Least They Can Do"
What started as Microbiological Associates grew to become part of several other, larger companies, including Whittaker Corp, BioWhittaker, Invitrogen, Cambrex, BioReliance, and Avista Capital Partners; for the profiles of those companies and others that sell HeLa, see *OneSource CorpTech Company Profiles* or Hoover.com.

For HeLa pricing information, search the product catalogs of any number of biomedical supply companies, including Invitrogen.com.

For patent information, search for HeLa in Patft.uspto.gov.

For information on the ATCC as a nonprofit, including financial statements, search for American Type Culture Collection on Guidestar.org; for its HeLa catalog entry, visit Atcc.org and search for HeLa.

For information on HeLa-plant hybrids, see "People-Plants," *Newsweek*, August 16, 1976; C. W. Jones, I. A. Mastrangelo, H. H. Smith, H. Z. Liu, and R. A. Meck, "Interkingdom Fusion Between Human (HeLa) Cells and Tobacco Hybrid (GGLL) Protoplasts," *Science*, July 30, 1976.

For an account of Dean Kraft's attempts to kill HeLa cells using "psychic healing," and thus cure cancer, see his book, *A Touch of Hope,* as well as related videos on YouTube.com (available by searching for Dean Kraft).

For the research done on the Lacks family's blood samples, see S. H. Hsu, B. Z. Schacter, et al., "Genetic Characteristics of the HeLa Cell," *Science* 191, no. 4225 (January 30, 1976). That research was funded by NIH Grant number 5P01GM019489-020025.

Chapter 25: "Who Told You You Could Sell My Spleen?"

Much of the Moore story appears in court and government documents, particularly the "Statement of John L. Moore Before the Subcommittee on Investigations and Oversight," House Committee on Science and Technology Hearings on the Use of Human Patient Materials in the Development of Commercial Biomedical Products, October 29, 1985; *John Moore v. The Regents of the University of California et al.* (249 Cal.Rptr. 494); and *John Moore v. The Regents of the University of California et al.* (51 Cal.3d 120, 793 P.2d 479, 271 Cal.Rptr. 146).

The Mo-cell patent is no. 4,438,032, available at Patft.uspto.gov.

The literature regarding the Moore trial and its implications is vast. Some useful sources include William J. Curran, "Scientific and Commercial Development of Human Cell Lines," *New England Journal of Medicine* 324, no. 14 (April 4, 1991); David W. Golde, "Correspondence: Commercial Development of Human Cell Lines," *New England Journal of Medicine,* June 13, 1991; G. Annas, "Outrageous Fortune: Selling Other People's Cells," *The Hastings Center Report* (November–December 1990); B. J. Trout, "Patent Law—A Patient Seeks a Portion of the Biotechnological Patent Profits in Moore v. Regents of the University of California," *Journal of Corporation Law* (Winter 1992); and G. B. White and K. W. O'Connor, "Rights, Duties and Commercial Interests: John Moore versus the Regents of the University of California," *Cancer Investigation* 8 (1990).

For a selection of media reports about the John Moore case, see Alan L. Otten, "Researchers' Use of Blood, Bodily Tissues Raises Questions About Sharing Profits," *Wall Street Journal,* January 29, 1996; "Court Rules Cells Are the Patient's Property," *Science,* August 1988; Judith Stone, "Cells for Sale," *Discover,* August 1988; Joan O'C. Hamilton, "Who Told You You Could Sell My Spleen?" *BusinessWeek,* April 3, 1990; "When Science Outruns Law," *Washington Post,* July 13, 1990; and M. Barinaga, "A Muted Victory for the Biotech Industry," *Science* 249, no. 4966 (July 20, 1990).

For the regulatory response to the Moore case, see "U.S. Congressional

Office of Technology Assessment, New Developments in Biotechnology: Ownership of Human Tissues and Cells—Special Report," Government Printing Office (March 1987); "Report on the Biotechnology Industry in the United States: Prepared for the U.S. Congressional Office of Technology Assessment," National Technical Information Service, U.S. Department of Commerce (May 1, 1987); and "Science, Technology and the Constitution," U.S. Congressional Office of Technology Assessment (September 1987). See also the never-passed "Life Patenting Moratorium Act of 1993," (103rd Congress, S.387) introduced February 18, 1993.

Details of the oil-consuming bacteria involved in Chakrabarty's lawsuit can be found in patent no. 4,259,444, available at Patft.uspto.gov. For more information on the lawsuit, see *Diamond v. Chakrabarty* (447 U.S. 303).

For further reading on other cell ownership cases mentioned in this chapter, see "Hayflick-NIH Settlement," *Science,* January 15, 1982; L. Hayflick, "A Novel Technique for Transforming the Theft of Mortal Human Cells into Praiseworthy Federal Policy," *Experimental Gerontology* 33, nos. 1–2 (January–March 1998); Marjorie Sun, "Scientists Settle Cell Line Dispute," *Science,* April 22, 1983; and Ivor Royston, "Cell Lines from Human Patients: Who Owns Them?" presented at the AFCR Public Policy Symposium, 42nd Annual Meeting, Washington, D.C., May 6, 1985; and *Miles Inc v. Scripps Clinic and Research Foundation et al.* (89-56302).

Chapter 26: Breach of Privacy

Whether the publication of a person's medical records would violate HIPAA today depends on many factors; most important, who released the records. HIPAA protects "all 'individually identifiable health information' . . . in any form or media, whether electronic, paper, or oral," but it only applies to "covered entities," which are health-care providers and health insurers that "furnish, bill or receive payment for" health care, and who transmit any covered health information electronically. This means any noncovered entity can release or publish a person's medical records without violating HIPAA.

According to Robert Gellman, a health-privacy expert who chaired a U.S. government subcommittee on privacy and confidentiality, any Hopkins faculty member releasing Henrietta's medical information today would most likely violate HIPAA, because Hopkins is a covered entity.

However, in October 2009, as this book went to press, portions of Henrietta's medical records were again published without her family's permission, this time in a paper coauthored by Brendan Lucey, of Michael O'Callaghan

Federal Hospital at Nellis Air Force Base; Walter A. Nelson-Rees, the HeLa contamination crusader who died two years before the article's publication; and Grover Hutchins, the director of autopsy services at Johns Hopkins. See B. P. Lucey, W. A. Nelson-Rees, and G. M. Hutchins, "Henrietta Lacks, HeLa Cells, and Culture Contamination," *Archives of Pathology and Laboratory Medicine* 133, no. 9 (September 2009).

Some of the information they published had previously appeared in Michael Gold's *Conspiracy of Cells*. They also published new information, including, for the first time, photos of her cervical biopsies.

According to Gellman, "It seems quite likely that HIPAA was violated in this case. But the only way to know for sure is an investigation that would go into complicated factors, including how they got the medical records in the first place." When I called Lucey, the paper's primary author, and asked how he'd gotten her records, and whether anyone had sought the family's permission to publish them, he told me the records had come from his co-author, Hutchins, at Hopkins. "Ideally, you'd like to get family approval," he said. "I believe Dr. Hutchins tried to track down a family member without success." The authors had obtained IRB approval to publish a series of articles using autopsy reports; in the other articles, they'd used initials to conceal patients' identities. Lucey pointed out that some of the information from Henrietta's medical records had been previously published, as had her name. "In this case protecting her identity with initials wouldn't have worked," he said. "Anyone can figure out who she is, since her name has already been connected with the cells."

When it comes to the dead and privacy: For the most part, the dead do not have the same right to privacy enjoyed by the living. One exception to that rule is HIPAA: "Even Thomas Jefferson's records, if they exist, are protected by HIPAA if they're held by a covered entity," Gellman said. "A hospital can't give away the records, regardless of whether the patient is dead or alive. Your right to privacy under HIPAA continues to exist until the sun runs out of hydrogen."

One other point to consider: though Henrietta was dead and therefore without the privacy rights of the living, many legal and privacy experts I talked with pointed out that the Lacks family could have argued that the release of Henrietta's medical records violated *their* privacy. There was no precedent for such a case at that time, but there have been such cases since.

For more information on the laws regarding confidentiality of medical records, and the debate surrounding them, see Lori Andrews's "Medical Genetics: A Legal Frontier"; *Confidentiality of Health Records* by Herman

Schuchman, Leila Foster, Sandra Nye, et al.; M. Siegler, "Confidentiality in Medicine: A Decrepit Concept," *New England Journal of Medicine* 307, no. 24 (December 9, 1982): 1518–1521; R. M. Gellman, "Prescribing Privacy," *North Carolina Law Review* 62, no. 255 (January 1984); "Report of Ad Hoc Committee on Privacy and Confidentiality," *American Statistician* 31, no. 2 (May 1977); C. Holden, "Health Records and Privacy: What Would Hippocrates Say?" *Science* 198, no. 4315 (October 28, 1977); and C. Levine, "Sharing Secrets: Health Records and Health Hazards," *The Hastings Center Report* 7, no. 6 (December 1977).

For related cases, see *Simonsen v. Swensen* (104 Neb. 224, 117 N.W. 831, 832, 1920); *Hague v. Williams* (37 N.J. 328, 181 A.2d 345. 1962); *Hammonds v. Aetna Casualty and Surety Co.* (243 F. Supp. 793 N.D. Ohio, 1965); *MacDonald v. Clinger* (84 A.D.2d 482, 446 N.Y.S.2d 801, 806); *Griffen v. Medical Society of State of New York* (11 N.Y.S.2d 109, 7 Misc. 2d 549. 1939); *Feeney v. Young* (191, A.D. 501, 181 N.Y.S. 481. 1920); *Doe v. Roe* (93 Misc. 2d 201, 400 N.Y.S.2d 668, 677. 1977); *Banks v. King Features Syndicate, Inc.* (30 F. Supp. 352. S.D.N.Y. 1939); *Bazemore v. Savannah Hospital* (171 Ga. 257, 155 S.E. 194. 1930); and *Barber v. Time* (348 Mo. 1199, 159 S.W.2d 291. 1942).

Chapter 27: The Secret of Immortality

For more on Jeremy Rifkin's lawsuits, see *Foundation on Economic Trends et al. v. Otis R. Bowen et al.* (No. 87-3393) and *Foundation on Economic Trends et al. v. Margaret M. Heckler, Secretary of the Department of Health & Human Services et al.* (756 F.2d 143). For media reports on the case, see Susan Okie, "Suit Filed Against Tests Using AIDS Virus Genes; Environmental Impact Studies Requested," *Washington Post,* December 16, 1987; and William Booth, "Of Mice, Oncogenes and Rifkin," *Science* 239, no. 4838 (January 22, 1988).

For the HeLa species debate, see L. Van Valen, "HeLa, a New Microbial Species," *Evolutionary Theory* 10, no. 2 (1991).

For more on cell immortality, see L. Hayflick and P. S. Moorhead, "The Serial Cultivation of Human Diploid Cell Strains," *Experimental Cell Research,* 25 (1961); L. Hayflick, "The Limited in Vitro Lifetime of Human Diploid Cell Strains," *Experimental Cell Research* 37 (1965); G. B. Morin, "The Human Telomere Terminal Transferase Enzyme Is a Ribonucleoprotein That Synthesizes TTAGGG Repeats," *Cell* 59 (1989); C. B. Harley, A. B. Futcher, and C. W. Greider, "Telomeres Shorten During Ageing of Human Fibroblasts," *Nature* 345 (May 31, 1990); C. W. Greider and E. H.

Blackburn, "Identification of Specific Telomere Terminal Transferase Activity in Tetrahymena Extracts," *Cell* 43 (December 1985).

For further reading on research into aging and human life extension, see Stephen S. Hall's *Merchants of Immortality.*

For a selection of HPV research involving HeLa cells, see Michael Boshart et al., "A New Type of Papillomavirus DNA, Its Presence in Genital Cancer Biopsies and in Cell Lines Derived from Cervical Cancer," *EMBO Journal* 3, no. 5 (1984); R. A. Jesudasan et al., "Rearrangement of Chromosome Band 11q13 in HeLa Cells," *Anticancer Research* 14 (1994); N. C. Popescu et al., "Integration Sites of Human Papillomavirus 18 DNA Sequences on HeLa Cell Chromosomes," *Cytogenetics and Cell Genetics* 44 (1987); and E. S. Srivatsan et al., "Loss of Heterozygosity for Alleles on Chromosome 11 in Cervical Carcinoma," *American Journal of Human Genetics* 49 (1991).

Chapter 28: After London

For HeLa symposium information, see notes for chapter 6.

For a sampling of Cofield's long legal history, see *Sir Keenan Kester Cofield v. ALA Public Service Commission et al.* (No. 89-7787); *United States of America v. Keenan Kester Cofield* (No. 91-5957); *Cofield v. the Henrietta Lacks Health History Foundation, Inc., et al.* (CV-97-33934); *United States of America v. Keenan Kester Cofield* (99-5437); and *Keenan Kester Cofield v. United States* (1:08-mc-00110-UNA).

Chapter 29: A Village of Henriettas

For the *Hopkins Magazine* story referenced here, see Rebecca Skloot, "Henrietta's Dance," *Johns Hopkins Magazine,* April 2000.

For other articles referenced in this chapter, see Rob Stepney, "Immortal, Divisible; Henrietta Lacks," *The Independent,* March 13, 1994; "Human, Plant Cells Fused: Walking Carrots Next?" *The Independent Record,* August 8, 1976 (via the *New York Times* news service); Bryan Silcock, "Man-Animal Cells Are Bred in Lab," *The* [London] *Sunday Times,* February 14, 1965; and Michael Forsyth, "The Immortal Woman," *Weekly World News,* June 3, 1997.

Chapter 31: Hela, Goddess of Death

The character named Hela appeared in many Marvel comic books. See, for example, "The Mighty Thor: The Icy Touch of Death!" *Marvel Comics Group* 1, no. 189 (June 1971).

Chapter 33: The Hospital for the Negro Insane

For the article describing Crownsville's history, see "Overcrowded Hospital 'Loses' Curable Patients," *Washington Post* (November 26, 1958). The history of Crownsville is also documented in "Maryland's Shame," a series by Howard M. Norton in the *Baltimore Sun* (January 9–19, 1949), and in material provided to me by Crownsville Hospital Center, including their "Historic Overview," "Census," and "Small Area Plan: Community Facilities."

A few years after Deborah and I visited Crownsville Hospital Center, it closed. For that story, see Robert Redding Jr., "Historic Mental Hospital Closes," *Washington Times* (June 28, 2004), available at Washingtontimes.com/news/2004/jun/28/20040628-115142-8297r/#at.

Chapter 36: Heavenly Bodies

The Bible given to me by Gary Lacks in this chapter was *Good News Bible: Today's English Version* (American Bible Society, 1992).

Afterword

The figures I cite on the number of Americans whose tissue is being used in research, as well as information on how that tissue is used, can be found in Elisa Eiseman and Susanne B. Haga's *Handbook of Human Tissue Sources.* For the National Bioethics Advisory Commission's investigation into the use of human tissues in research, and its policy recommendations, see *Research Involving Human Biological Materials: Ethical Issues and Policy Guidance,* vol. 1: *Report and Recommendations of the National Bioethics Advisory Commission,* and vol. 2: *Commissioned Papers* (1999).

The literature on the use of human tissues in research, and the ethical and policy debate surrounding it, is vast and includes E. W. Clayton, K. K. Steinberg, et al., "Informed Consent for Genetic Research on Stored Tissue Samples," *Journal of the American Medical Association* 274, no. 22 (December 13, 1995): 1806–7, and resulting letters to the editor; *The Stored Tissue Issue: Biomedical Research, Ethics, and Law in the Era of Genomic Medicine,* by Robert F. Weir and Robert S. Olick; *Stored Tissue Samples: Ethical, Legal, and Public Policy Implications,* edited by Robert F. Weir; *Body Parts: Property Rights and the Ownership of Human Biological Materials,* by E. Richard Gold; *Who Owns Life?,* edited by David Magnus, Arthur Caplan, and Glenn McGee; and *Body Bazaar,* by Lori Andrews.

For a selection of related lawsuits, see *Margaret Cramer Green v. Commissioner of Internal Revenue* (74 T.C. 1229); *United States of America v. Dorothy R. Garber* (78-5024); *Greenberg v. Miami Children's Hospital Research Institute* (264 F.Supp.2d 1064); *Steven York v. Howard W.*

Jones et al. (89-373-N); *The Washington University v. William J. Catalona, M.D., et al.* (CV-01065 and 06-2301); *Tilousi v. Arizona State University Board of Regents* (04-CV-1290); *Metabolite Laboratories, Inc., and Competitive Technologies, Inc., v. Laboratory Corporation of America Holdings* (03-1120); *Association for Molecular Pathology et al. v. United States Patent and Trademark Office; Myriad Genetics et al.* (case documents online at aclu.org/brca/); and *Bearder et al. v. State of Minnesota and MDH* (complaint online at cchfreedom.org/pr/pro31109.php).

Reading Group Guide

For nearly sixty years HeLa cells have been essential to scientific research, but who was the woman from whom these cells were taken? This question prompted Rebecca Skloot to look at every facet of the life of Henrietta Lacks, and this book is the answer to that question.

Skloot's account of her ten-year search for the details of Henrietta's life and legacy takes her from a poor black neighborhood outside Baltimore where she meets Henrietta's family, to Johns Hopkins where Henrietta was treated for cancer and where her immortal cells were first cultured, and then to a small Virginia town and tobacco plantation to see where Henrietta grew up. Along the way she spends time with Henrietta's daughter Deborah who joins Skloot on her mission to find out more about the proud woman who has stared at her through a photograph for nearly a decade.

This guide is designed to enhance your reading group's discussion of this groundbreaking book.

Questions for Discussion

1. On page xiii, Rebecca Skloot states "This is a work of nonfiction. No names have been changed, no characters invented, no events fabricated." Consider the process Skloot went through to verify dialogue, re-create scenes, and establish facts. Imagine trying to re-create scenes such as when Henrietta discovered her tumor (page 15). What does Skloot say on pages xiii–xiv and in the notes section (page 346) about how she did this?

2. One of Henrietta's relatives said to Skloot, "If you pretty up how people spoke and change the things they said, that's dishonest" (page xiii). Throughout, Skloot is true to the dialect in which people spoke to her: the Lackses speak in a heavy Southern accent, and Lengauer and Hsu speak as non-native English speakers. What impact did the decision to maintain speech authenticity have on the story?

3. As much as this book is about Henrietta Lacks, it is also about Deborah learning of the mother she barely knew, while also finding out the truth about her sister, Elsie. Imagine discovering similar information about one of your family members. How would you react? What questions would you ask?

4. In a review for the *New York Times*, Dwight Garner writes, "Ms.

Skloot is a memorable character herself. She never intrudes on the narrative, but she takes us along with her on her reporting." How would the story have been different if she had not been a part of it? What do you think would have happened to scenes like the faith healing on page 289? Are there other scenes you can think of where her presence made a difference? Why do you think she decided to include herself in the story?

5. Deborah shares her mother's medical records with Skloot, but is adamant that she not copy everything. On page 284 Deborah says, "Everybody in the world got her cells, only thing we got of our mother is just them records and her Bible." Discuss the deeper meaning behind this sentence. Think not only of her words, but also of the physical reaction she was having to delving into her mother's and sister's medical histories. If you were in Deborah's situation, how would you react to someone wanting to look into your mother's medical records?

6. This is a story with many layers. Though it's not told chronologically, it is divided into three sections. Discuss the significance of the titles given to each part: Life, Death, and Immortality. How would the story have been different if it were told chronologically?

7. As a journalist, Skloot is careful to present the encounter between the Lacks family and the world of medicine without taking sides. Since readers bring their own experiences and opinions to the text, some may feel she took the scientists' side, while others may feel she took the family's side. What are your feelings about this? Does your opinion fall on one side or the other, or somewhere in the middle, and why?

8. Henrietta signed a consent form that said, "I hereby give consent to the staff of The Johns Hopkins Hospital to perform any operative procedures and under any anaesthetic either local or general that they may deem necessary in the proper surgical care and treatment of: _____ " (page 31). Based on this statement, do you believe TeLinde and Gey had the right to obtain a sample from her cervix to use in their research? What information would they have had to give her for Henrietta to give *informed* consent? Do you think Henrietta would have given explicit consent to have a tissue sample used in medical research if she had been given all the information? Do you always thoroughly read consent forms before signing them?

9. In 1976, when Mike Rogers's *Rolling Stone* article was printed, many viewed it as a story about race (see page 197 for reference). How do you think public interpretation might have been different if the piece had been published at the time of Henrietta's death in 1951? How is this different from the way her story is being interpreted today? How do you think

Henrietta's experiences with the medical system would have been different had she been a white woman? What about Elsie's fate?

10. Consider Deborah's comment on page 276: "Like I'm always telling my brothers, if you gonna go into history, you can't do it with a hate attitude. You got to remember, times was different." Is it possible to approach history from an objective point of view? If so, how and why is this important, especially in the context of Henrietta's story?

11. Deborah says, "But I always have thought it was strange, if our mother cells done so much for medicine, how come her family can't afford to see no doctors? Don't make no sense" (page 9). Should the family be financially compensated for the HeLa cells? If so, who do you believe that money should come from? Do you feel the Lackses deserve health insurance even though they can't afford it? How would you respond if you were in their situation?

12. Dr. McKusick directed Susan Hsu to contact Henrietta's children for blood samples to further HeLa research; neither McKusick nor Hsu tried to get informed consent for this research. Discuss whether or not you feel this request was ethical. Further, think about John Moore and the patent that had been filed without his consent on his cells called "Mo" (page 201). How do you feel about the Supreme Court of California ruling that states when tissues are removed from your body, with or without your consent, any claim you might have had to owning them vanishes?

13. Religious faith and scientific understanding, while often at odds with each other, play important roles in the lives of the Lacks family. How does religious faith help frame the Lacks' response to and interpretation of the scientific information they receive about HeLa? How does Skloot's attitude towards religious faith and science evolve as a result of her relationship with the Lackses?

14. On page 261, Deborah and Zakariyya visit Lengauer's lab and see the cells for the first time. How is their interaction with Lengauer different from the previous interactions the family had with representatives of Johns Hopkins? Why do you think it is so different? What does the way Deborah and Zakariyya interact with their mother's cells tell you about their feelings for her?

15. Reflect upon Henrietta's life: What challenges did she and her family face? What do you think their greatest strengths were? Consider the progression of Henrietta's cancer in the last eight months between her diagnosis and death. How did she face death? What do you think that says about the type of person she was?

About the Author

REBECCA SKLOOT is an award-winning science writer whose work has appeared in *The New York Times Magazine; O, The Oprah Magazine; Discover;* and many others. She is coeditor of *The Best American Science Writing 2011* and has worked as a correspondent for NPR's *Radiolab* and PBS's Nova *ScienceNOW.* She was named one of five surprising leaders of 2010 by the *Washington Post.* Skloot's debut book, *The Immortal Life of Henrietta Lacks,* took more than a decade to research and write, and instantly became a *New York Times* bestseller. It was chosen as a best book of 2010 by more than sixty media outlets, including *Entertainment Weekly, People,* and the *New York Times.* It is being translated into more than twenty-five languages, adapted into a young reader edition, and being made into an HBO film produced by Oprah Winfrey and Alan Ball. Skloot is the founder and president of The Henrietta Lacks Foundation. She has a B.S. in biological sciences and an MFA in creative nonfiction. She has taught creative writing and science journalism at the University of Memphis, the University of Pittsburgh, and New York University. She lives in Chicago. For more information, visit her website at RebeccaSkloot.com, where you'll find links to follow her on Twitter and Facebook.